FORTSCHRITTE DER PHYSIK

HERAUSGEGEBEN IM AUFTRAGE DER PHYSIKALISCHEN GESELLSCHAFT
DER
DEUTSCHEN DEMOKRATISCHEN REPUBLIK

VON F. KASCHLUHN, A. LÖSCHE, R. RITSCHL UND R. ROMPE

HEFT 4 · 1977 BAND 25

THEORIA CUM PRAXI

AKADEMIE VERLAG BERLIN

EVP 10,– M
31728

BEZUGSMÖGLICHKEITEN

Bestellungen sind zu richten

— **in der DDR** an eine Buchhandlung oder an den Akademie-Verlag, DDR-108 Berlin, Leipziger Straße 3—4

— **im sozialistischen Ausland** an eine Buchhandlung für fremdsprachige Literatur oder an den zuständigen Postzeitungsvertrieb

— **in der BRD und Westberlin** an eine Buchhandlung oder an die Auslieferungsstelle KUNST UND WISSEN, Erich Bieber, 7 Stuttgart 1, Wilhelmstraße 4—6

— **in Österreich** an den Globus-Buchvertrieb, 1201 Wien, Höchstädtplatz 3

— **im übrigen Ausland** an den Internationalen Buch- und Zeitschriftenhandel; den Buchexport, Volkseigener Außenhandelsbetrieb der Deutschen Demokratischen Republik, DDR-701 Leipzig, Postfach 160, oder

an den Akademie-Verlag, DDR-108 Berlin, Leipziger Straße 3—4

Zeitschrift „Fortschritte der Physik"

Herausgeber: Prof. Dr. Frank Kaschluhn, Prof. Dr. Artur Lösche, Prof. Dr. Rudolf Ritschl, Prof. Dr. Robert Rompe, im Auftrag der Physikalischen Gesellschaft der Deutschen Demokratischen Republik.
Verlag: Akademie-Verlag, DDR-108 Berlin, Leipziger Straße 3—4; Fernruf: 2200441; Telex-Nr. 114420; Postscheckkonto: Berlin 35021; Bank: Staatsbank der DDR, Berlin, Konto-Nr.: 6836-26-20712.
Chefredakteur: Dr. Lutz Rothkirch.
Anschrift der Redaktion: Sektion Physik der Humboldt-Universität zu Berlin, DDR-104 Berlin, Hessische Straße 2.
Veröffentlicht unter der Lizenznummer 1324 des Presseamtes beim Vorsitzenden des Ministerrates der Deutschen Demokratischen Republik.
Gesamtherstellung: VEB Druckhaus „Maxim Gorki", DDR-74 Altenburg, Carl-von-Ossietzky-Straße 30/31.
Erscheinungsweise: Die Zeitschrift „Fortschritte der Physik" erscheint monatlich. Die 12 Hefte eines Jahres bilden einen Band.
Bezugspreis je Band: 180,— M zuzüglich Versandspesen (Preis für die DDR: 120,— M). Preis je Heft 15,— M (Preis für die DDR: 10,— M).
Bestellnummer dieses Heftes: 1027/25/4.

AN (EDV) 57618

Fortschritte der Physik 25, 203–271 (1977)

A Survey on Dual Tree and Loop Amplitudes

D. Ebert

Joint Institute for Nuclear Research, Dubna, USSR

and

H.-J. Otto

Sektion Physik der Humboldt-Universität zu Berlin, GDR

A Survey on Dual Tree and Loop Amplitudes

I. Introduction

Since in 1968 Veneziano first wrote down a dual scattering amplitude in terms of Euler's beta function [1], the dual resonance model (DRM) was subjected to a rapid and successful development. This may easily be understood by the fact that the model incorporates many attractive features such as Lorentz-invariance, crossing-symmetry, Regge behaviour and selfconsistent factorization properties that make it a very promising candidate for describing at least some general properties of strong interaction scattering amplitudes. From the analyticity point of view the meromorphic dual amplitudes with an infinity of narrow resonances on linear trajectories $\alpha(s) = a + \alpha' s$ have to be interpreted as Born ('tree') terms of some generalized perturbation theory. Higher order loop contributions can be constructed and deliver unitarity corrections including imaginary parts of the originally real Regge trajectories. This way, the loops transform the infinitely narrow resonances of the Born term into unstable particles with a finite width. A large part of this paper will be devoted to these unitarity corrections. By virtue of an underlying Feynman-like diagram structure the dual resonance model bears a strong resemblance to usual field theory. This can even be made more explicit when considering the limit of the model with the slope parameter α' of the input Regge trajectories going to zero [2]. But there are some specific features not encountered in other models. This concerns, for example, the character of divergences of loop expressions due to the exponentially growing mass spectrum as well as the natural appearance of a new singularity in vacuum channels of a certain class of loop diagrams that can be interpreted as the exchange of a 'Pomeron' trajectory. Unfortunately, one has to fix the intercept of the input trajectory and the space-time dimensionality D to unphysical values ($a = 1$, $D = 26$) to make this new Pomeron singularity a Regge pole and prevent it from being a unitarity violating cut. The intercept of the Pomeron trajectory then is $\alpha_p(0) = 2$, i.e. it is by the same factor 2 too large as the intercept of the leading meson trajectory ($a = 1$). The necessity for choosing this value $a = 1$ implying a tachyon ground state at mass $m_0 = (-\alpha')^{-1/2}$ lies in the fact that this is just the prerequisite for having an

infinite gauge algebra at work eliminating all negative norm (ghost) states. Furthermore, the original dual resonance model neither contained fermions nor half integer spaced trajectories ($\pi - \rho$) found in nature. This problem could be overcome by the more sophisticated models of NEVEU and SCHWARZ [3] for bosons and of RAMOND [4] for fermions that soon were understood to be two sectors of one and the same (NSR-) model. The critical space-time dimensionality enforced by the Pomeron is lowered by the additional structure introduced, but it remains an unphysical one ($D = 10$).
Meanwhile, the underlying space-time description of the dual theory was found to be given by the quantum mechanics of one-dimensionally extended objects without rest mass or, in short, by the theory of quantized strings [5]. The transition amplitudes between different multi-string states just come out to be the dual ones [6], where the NSR-modification corresponds to additional spin distributions along the string [7].
The main aim of this review is to present some essential ideas and result concerning both the single- and multi-loop contributions in dual scattering amplitudes. In dealing with the very involved multi-loop expression it is more in our intentions to give some examples for handling them in practical applications than reiterating their derivation at length. To do these things on a well-defined basis and to make this article as selfcontained as possible, we have added an introduction into general structures and the operator formulation of the dual Born terms in different dual models. The reader being intersted in a more detailed representation of dual tree amplitudes including the dual string theory can be referred to quite a number of comprehensive reviews on this field [8–12]. In closing our first part on dual tree amplitudes we give also some examples of using them phenomenologically.
In chapter III we first discuss the fundamental one-loop operators on the basis of which then all relevant multi-loop expressions will be constructed. The alternative formulation in terms of Abelian integrals reflecting automorphic structures on Riemann surfaces will be sketched afterwards. To obtain explicit functions for unitarity corrections to the trajectory functions we then discuss the high energy behaviour and the factorization properties of the asymptotic expressions. The singularities of the nonplanar orientable two-loop diagram will be exposed as an example for the application of the representation by Abelian integrals. Finally, we briefly review the main ideas of the renormalization procedure in dual models.

II. Formulation of the Dual Theory — Tree Approximation

II.1. The Generalized Veneziano Model (GVM)

II.1.1. The Veneziano ansatz

In 1968 Veneziano proposed a scattering amplitude for the kinematically simple process $\pi\pi \to \pi\omega$ in terms of a beta function [1]. For the elastic scattering of two scalar particles the Veneziano model is equivalent to the following crossing symmetric amplitude

$$T(s, t, u) = g^2\big[B\big(-\alpha(s), -\alpha(t)\big) + B\big(-\alpha(t), -\alpha(u)\big) + B\big(-\alpha(s), -\alpha(u)\big)\big]. \qquad \text{(II.1)}$$

Here the beta function is given by the integral representation

$$B\big(-\alpha(s), -\alpha(t)\big) = \int_0^1 dx\, x^{-\alpha(s)-1}(1-x)^{-\alpha(t)-1}; \qquad \text{(II.2)}$$

s, t, u are the usual Mandelstam invariants (comp. fig. 1)

$$\left(s = (p_1 + p_2)^2, \quad t = (p_2 + p_3)^2, \quad u = (p_2 + p_4)^2; \quad \sum_{i=1}^{4} p_i = 0; \quad s + t + u = 4m_0^2\right)$$

and g is an universal coupling constant.
The function $\alpha(s_{ij})$ $\left(s_{ij} = (p_i + p_j)^2\right)$ will prove out to be the leading Regge trajectory in all three channels (i, j) and is put into the ansatz (II.1) as a real and linear rising function of s_{ij}

$$\alpha(s_{ij}) = a + \alpha' s_{ij} = a + \frac{1}{2} s_{ij}. \tag{II.3}$$

By the last equation we have taken the convention to measure each momentum in units $(2\alpha')^{-1/2}$ (we use generally the units $\hbar = c = 1$). It is further implied that the external scalar particles lie on the same trajectory $\alpha(s)$, i.e. for $p_i^2 = m_0^2 (i = 1, \dots 4)$ we have $\alpha(p_i^2) = a + 1/2\, m_0^2 = 0$. The intercept and slope parameters a, α' of the trajectory function and the coupling constant g are the only free parameters of the model.

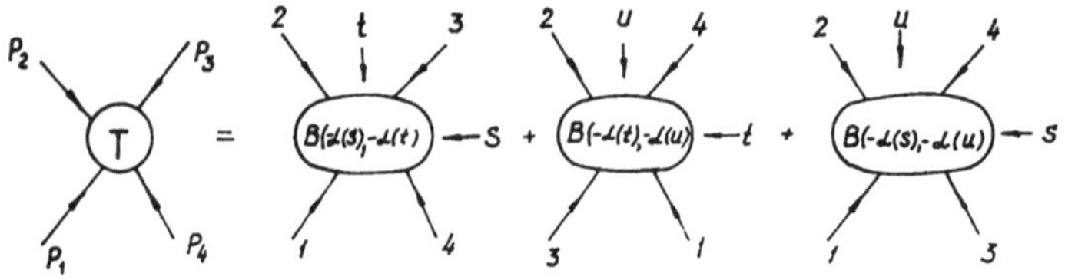

Fig. 1. The noncyclic configurations of the four-point amplitude

Using the series expansion for an individual beta-function in eq. (II.1) one obtains

$$B\left(-\alpha(s), -\alpha(t)\right) = \sum_{l=0}^{\infty} \frac{1}{l - \alpha(s)} \frac{1}{l!} \frac{\Gamma\left(l + \alpha(t) + 1\right)}{\Gamma\left(\alpha(t) + 1\right)} \tag{II.4a}$$

$$= \sum_{l=0}^{\infty} \frac{1}{l - \alpha(t)} \frac{1}{l!} \frac{\Gamma\left(l + \alpha(s) + 1\right)}{\Gamma\left(\alpha(s) + 1\right)}. \tag{II.4b}$$

The equality of the expansion of an individual Veneziano function into a series of s-channel or t-channel resonances expresses in this most simple case the so called duality property of the model. Each of the three beta-functions in eq. (II.1) can be brought into a one-to-one correspondence to a definite arrangement of the external particles. All cyclic or anticyclic permutations of a given order are considered to be identical (fig. 1). Therefore any Veneziano term is represented by a planar 'tree' diagram, and different terms can be transformed into one another by a nonplanar operation (noncyclic particle interchange) only. Each of these planar diagrams shares the duality property of eq. (II.4) with respect to the corresponding pair of planar variables $s_{ij} = s_{kl}$ and $s_{il} = s_{jk}((ijkl) =$ permutation of (1234); comp. fig. 2a)
In this sense the Veneziano model is referred to as a planar dual model.[1]) Contrary to usual Feynman rules the two diagrams shown in fig. 2a may not be added, as they

[1]) We only mention here the existence of the nonplanar extension of the dual model by Shapiro and Virasoro [*14, 9*]. Though the splitting up into cyclic symmetric planar terms cannot be made there, the amplitude as a whole is again crossing symmetric containing poles in each Mandelstam channel.

represent the same function in 'dualized' configurations. The two channels s_{ij} and s_{jk} are said to be dual to each other.
The attractiveness of the Veneziano model rests primarily in the fact that it accounts quite for a large set of basic principles believed to be essential for an S-matrix theory [13] at least of strong interactions:

(i) The model is crossing symmetric and Lorentz invariant by construction.
(ii) It is analytic in the sense that the only singularities in any invariant mass squared variable of a scattering channel are poles. (comp. eq. (II.4)).
(iii) It is Regge behaved in any direction of the complex (e.g.) s-plane outside an arbitrarily narrow wedge containing the real axis

$$T(s, t, u) \underset{\substack{|s|\to\infty, t \text{ fixed} \\ |\arg s| \geq \varepsilon > 0}}{\sim} g^2 \Gamma(-\alpha(t)) [\alpha(s)]^{\alpha(t)} [1 + e^{-i\pi\alpha(t)}] \tag{II.5}$$

(only the first two beta-functions on eq. (II.1) contribute to the signaturized Regge behaviour (II.5), the (s, u)-term vanishes exponentially.)

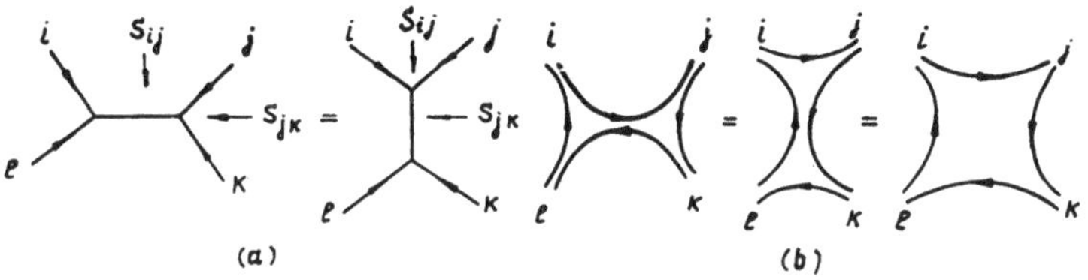

Fig. 2. The duality property of the four-point amplitude represented by Feynman-like diagrams (a) and quark-like diagrams (b)

From eq. (II.4a) we see that the residuum at the pole $l = \alpha(s_l)$ of the beta function is a polynomial of degree l in t due to the linearity of the trajectory. Thus, the Veneziano term $B(-\alpha(s), -\alpha(t))$ is made up totally from an infinity of narrow ($\Gamma_{\text{res}} \sim \text{Im}\,\alpha = 0$) and equally spaced resonances in the s-(or t-) channel with spins $0 \leqq j \leqq l$ and masses $s_l = m_l{}^2 = 2(l - a)$. The splitting up of each level s_l into a contribution with the leading spin $j = l$ and the daughters $0 \leqq j \leqq l - 1$ corresponds to an expansion of the residuum in terms of Legendre polynomials (0 (3) reduction). As will become manifest soon, the degeneracy at each level is, generally speaking, much larger than expected from 0(3) reduction alone.
The necessity of excluding the real s-axis in the asymptotic expression (II.5) is, of course, related to the presence of the infinite number of stable particle poles there, or, in other words, to the narrow width approximation expressed by the reality of the trajectory. It is obvious that this approximation has to be improved so that resonances of finite width lying on complex trajectories can emerge[2]). Thereby cuts should appear in the up to now purely meromorphic amplitude so that unitarity can be fulfilled. We shall come back to this point in chapt. III where the question of higher order corrections to the Veneziano model will be investigated.
We remark here that the Veneziano model fits nicely into the earlier developed scheme of Harari-Rosner duality diagrams being founded on the lack of exotic resonances [15]. All external mesons are represented there by a pair of quark-anti-quark lines ($q\bar{q}$-states). A legal diagram does not contain crossed quark lines or external mesons built up only by one and the same line. The quark picture of fig. 2b illustrates that the two dual represen-

[2]) One cannot simply replace the linear trajectories by complex ones, as in this case there would be at any level $m_l{}^2$ an infinity of 'ancestor' states with arbitrarily high spin j.

tations of a single Veneziano term merely correspond to different ways of stretching the duality diagram. The analogue model of Fairlie and Nielsen [16] and the string picture [6] lend some more physical support to this structure.

II.1.2. N-point generalization

Let us rewrite the Veneziano function (II.2) in a form which is suitable for further generalizations. For this purpose one attributes an integration variable u_{ij} to every planar channel s_{ij} of a given diagram (e.g. $s_{12} = s$, $s_{23} = t$; $\alpha_{ij} \equiv \alpha(s_{ij})$) and imposes the duality constraint

$$u_{12} + u_{23} = 1 \tag{II.6}$$

on the integration region. The integral (II.2) is then recast into the symmetric form

$$B(-\alpha_{12}, -\alpha_{23}) = \int_0^1\!\!\int \left\{ \prod_{\substack{(i,j)= \\ (1,2),(2,3)}} du_{ij} u_{ij}^{-\alpha_{ij}-1} \right\} \delta(u_{12} + u_{23} - 1). \tag{II.7}$$

In discussing the N-point generalization of the Veneziano model, i.e. the dual model for N external scalar particles of mass m_0, we shall limit ourselves to a fixed ordering $1, 2, \ldots, N$ of the external particles. The summation over the $1/2 \cdot (N-1)!$ inequivalent permutations producing a crossing symmetric total amplitude is always understood.

For a dual N-point (tree) function B_N $(1, 2, \ldots, N)$ (fig. 3a) there are generally $1/2 \cdot N \times (N-3) = K_N$ planar channels with invariant mass variables $s_{ij} = (p_i + p_{i+1} + \cdots + p_{j-1} + p_j)^2$ ($i, j \bmod N$). They correspond to all the chords drawn into the N-sided polygon of fig. (3c). As there are only $N-3$ non-intersecting chords there are only

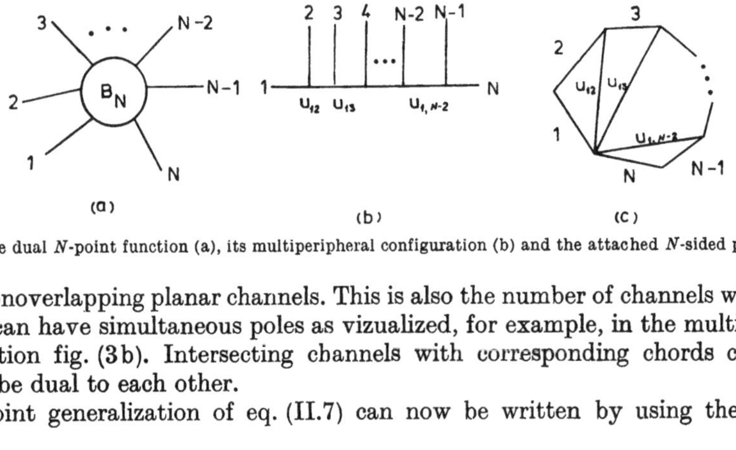

Fig. 3. The dual N-point function (a), its multiperipheral configuration (b) and the attached N-sided polygon (c).

$N-3$ nonoverlapping planar channels. This is also the number of channels where a tree diagram can have simultaneous poles as vizualized, for example, in the multiperipheral configuration fig. (3b). Intersecting channels with corresponding chords crossed are called to be dual to each other.

The N-point generalization of eq. (II.7) can now be written by using the following rules [17]

(i) Write a cyclic symmetric integrand in terms of K_N 'Chan-variables' u_{ij} inserted for each planar channel and ranging between 0 and 1.

(ii) Impose $K_N - (N-3)$ 'duality constraints' (Comp. eq. II.6)

$$u_{ij} = 1 - \prod_{(kl)^{ij}} u_{kl} \tag{II.8}$$

($(kl)^{ij}$ denotes the channels dual to the channel (ij)). These constraints ensure that the vanishing of the variable leading to poles in the (ij)-channel excludes automatically the simultaneous vanishing of the dual variables u_{kl} and therefore simultaneous poles in dual channels.

(iii) The measure function in the integral has to be specified in such a way that a cyclic change in the choice of the set of $N-3$ independent non dual integration variables preserves the functional form of the integrand.

Choosing, for example, the $N-3$ 'multiperipheral variables' $u_{12}, u_{13}, \ldots, u_{1,N-2}$ (see fig. 3b) as the independent ones one may express the dependent ones, using eq. (II.8), as

$$u_{ij} = \frac{(1-x_{i,j})(1-x_{i-1,j+1})}{(1-x_{i-1,j})(1-x_{i,j+1})} \quad (1 < i < j < N) \tag{II.9}$$

$$(x_{i,j} \equiv u_{1i} u_{1,i+1} \cdot u_{1,j+2} \ldots u_{1,j-1} \qquad u_{11} = u_{1N} = 0).$$

This way Bardakci and Ruegg obtained the following integral representation for the N-point function[3]) [17]

$$B_N(1, 2, \ldots, N) = g^{N-2} \int_0^1 \cdots \int_0^1 \left(\prod_{j=2}^{N-2} du_{ij} u_{1j}^{-\alpha_{ij}-1} (1-u_{1j})^{a-1} \right)$$

$$\times \prod_{1<k<l<N} (1 - u_{1k} u_{1,k+1} \ldots u_{1,l-1})^{-p_k p_l}. \tag{II.10}$$

The N-point generalization (II.10) of the beta function defines the so called generalized Veneziano model (GVM).

II.1.3. The Koba-Nielsen form

Though the integral representation (II.10) gives a cyclic symmetric B_N-function by construction, the complexity of the integrand due to the quite involved solutions of the constraint equations (II.8) makes a simplification desirable. Koba and Nielsen (KN) were successful in transforming the integral representation for B_N into a more manageable form [18]. The essential point in the KN-representation is, that there appear from the beginning only N (KN-) variables which are in one-to-one correspondence to the external particles. The integrand is invariant under any simultaneous projective transformation of all the KN-variables. As these transformations form a representation of a 3-parameter group, three variables can arbitrarily be fixed and one is left with $N-3$ independent integrations as required by duality.

To make this plausible with the Bardakci-Ruegg representation (II.10) as a starting point, one remembers, that the dependent variables u_{ij} (eq. (II.9) are anharmonic ratios which are, as is well known, projective invariants. One easily casts eq. (II.9) into the following form

$$u_{ij} = \frac{z_i - z_j}{z_i - z_{j+1}} \frac{z_{i-1} - z_{j+1}}{z_{i-1} - z_j} \equiv (z_i, z_{i-1}, z_j, z_{j+1}) \tag{II.11}$$

where

$$z_i = \begin{cases} x_{2i} = u_{12} u_{13} \ldots u_{1,i-1} & \text{for } 2 < i < N \\ +\infty & \text{for } i = 1 \\ 1 & \text{for } i = 2 \\ 0 & \text{for } i = N \end{cases} \tag{II.12a}$$

$$0 = z_N \leqq z_{N-1} \leqq \cdots \leqq z_3 \leqq z_2 = 1 < z_1 = \infty. \tag{II.12b}$$

The KN-variables z_i lie on the real axis in very the same cyclic order as the external particles 1, 2, ..., N and can therefore be associated to the latter. We consider now an

[3]) From now on we multiply the Veneziano term with the corresponding power of the coupling constant g.

arbitrary projective (Möbius) transformation of the form

$$z' = \frac{az + b}{cz + d} = \gamma(z). \tag{II.13}$$

The transformation $\gamma(t)$ is obviously given by the 2×2 matrix

$$p = \begin{pmatrix} a & b \\ c & d \end{pmatrix} \tag{II.14}$$

and two successive projective transformations just correspond to the product of the respective 2×2 matrices. To guarantee the group property one has to restrict to matrices with nonvanishing determinant, e.g. $\det p > 0$.
This group of transformations $\gamma(z)$ is then isomorphic to $SU(1, 1)$ (complex p_{ij}: $p_{11} = p_{22}^*$, $p_{12} = p_{21}^*$; $\det p = 1$) or to $SL(2, R)$ (real p_{ij}, $\det p = 1$), and they all are isomorphic to the group of Lorentz-transformations in a space with two space and one time dimension, i.e. $0(2, 1)$. It is a three parameter group, so that any group element γ is fixed with the given transformation of three points. It is sometimes convenient not to use the 2×2 matrix p, but a symbol introduced by Lovelace [19] as representation containing three points z and their images $\gamma(z)$

$$\gamma = \begin{bmatrix} z_1 & z_2 & z_3 \\ \gamma(z_1) & \gamma(z_2) & \gamma(z_3) \end{bmatrix}. \tag{II.15}$$

Due to the fact that the Bardakci-Ruegg formula (II.10) contains only the z's themselves ($u_{1,j} = z_{j+1}/z_j$, $2 \leqq j \leqq N - 2$) and the anharmonic rations (II.11) one obtains the following representation ($z_{N+1} = z_1$) [18]

$$B_N = g^{N-2} \lim_{\substack{z_1 \to \infty \\ z_2 \to 1 \\ z_N \to 0}} \int\limits_{z_1 > z_2 > \quad > z_N} [d^3 V_{12N}]^{-1} \left[\prod_{i=1}^{N} dz_i |z_i - z_{i+1}|^{a-1} \right] \prod_{1 \leqq k < l \leqq N} |z_k - z_l|^{-p_k p_l} \tag{II.16}$$

where the formal measure

$$[d^3 V_{12N}]^{-1} = \left[\frac{dz_1\, dz_2\, dz_N}{|z_1 - z_2|\, |z_1 - z_N|\, |z_2 - z_N|} \right]^{-1} \tag{II.17}$$

essentially undoes the three redundant integrations. Koba and Nielsen proved the Möbius invariance of the total integrand including the measure so that three arbitrary variables among the N KN-variables may indeed be fixed at arbitrary values.
The real axis can be furthermore transformed projectively into the unit circle in the z-plane, where one has now to fix three complex points (z_a, z_b, z_c) arbitrarily on that circle. The invariance group is then $SU(1,1)$ instead of $SL(2, R)$ in the real KN-representation. Therefore the KN-representation can be written generally as follows:

$$B_N(1, 2, \ldots, N) = g^{N-2} \oint\limits_{z_1 \to z_2 \to \cdot \quad \to z_N \to z_1} \frac{1}{d^3 V_{abc}} \left(\prod_{i=1}^{N} dz_i |z_i - z_{i+1}|^{a-1} \right) \prod_{1 \leqq k < l \leqq N} |z_k - z_l|^{-p_k p_l}. \tag{II.16'}$$

The symbol at the integral merely indicates the imposed cyclic ordering condition for the KN-variables, varying on the chosen KN-circle. The poles of the B_N-function appear here due to the end point singularities of the integration region associated with the coincidence of at least two KN-variables, say, $z_i = z_j$.
The representation (II.16') shows very clearly the cyclic invariance of the model, as any cyclic change of the momenta can be compensated by a cyclic interchange of the Z-variables. It should be mentioned that the derivation of the KN-form of the dual

amplitudes was an essential step for disclosing the group theoretical background of duality. The intimate connection of projective invariance and duality proved out later to be essential for the construction and understanding both of higher order loop graphs and of more sophisticated dual models, as will be discussed later.

II.1.4. Operator formulation

The spectrum of the Veneziano amplitudes consisting of equally spaced intermediate resonances suggests an underlying harmonic oscillator structure. This point was made explicit by NAMBU and FUBINI, GORDON and VENEZIANO in the early times of the dual theory already [20]. One introduces an infinite set of creation and destruction oscillator operators $a_n^{+\mu}$, a_n^{μ} in such a way that the GVM-tree amplitudes emerge as the vacuum expectation value of certain functions of the oscillator operators [21].
The relevant nonvanishing commutator in the algebra of the harmonic oscillator operators reads
($g_{\mu\nu} = \text{diag.}\,(1, -1, -1, -1)$)

$$[a_m^{\mu}, a_n^{+\nu}] = -g^{\mu\nu}\delta_{mn}(m, n = 1, 2, \ldots). \tag{II.18}$$

Here, the Lorentz metric $g^{\mu\nu}$ guarantees covariance, but it instantaneously introduces negative norm (ghost) states in the corresponding Fock space as is the case in any relativistic theory with vectorial oscillators. We shall come back to this point later.
Moreover, it is convenient to introduce sime 'zero mode' coordinate and momentum operators x_μ, p_ν with the canonical commutation relations[4])

$$[x_\mu, p_\nu] = -ig_{\mu\nu} \tag{II.19}$$

and a vacuum state $|0; 0\rangle$ which is the zero eigenstate both for p_μ and all the destruction operators $a_{n,\mu}$

$$p_\mu|0; 0\rangle = a_{n,\mu}\,|0; 0\rangle = 0. \tag{II.20}$$

Let us introduce further a vertex operator for the absorption (or emission) of a scalar particle of momentum p with the KN variable z attributed to this particle

$$V_0(p, z) = g\!:\!e^{-ipQ(z)}\!: = ge^{-ipQ^+(z)}e^{-ipQ^0(z)}e^{-ipQ^-(z)} \tag{II.21}$$

where $Q_\mu(z)$ is a so called FUBINI-VENEZIANO field [21]

$$\begin{aligned} Q_\mu(z) &= Q_\mu{}^0(z) + Q_\mu{}^+(z) + Q_\mu{}^-(z) \\ &= [x_\mu - i\ln z\cdot p_\mu] + \left[-i\sum_{m=1}^{\infty}\frac{1}{\sqrt{m}}\,a^+_{m,\mu}z^m\right] + \left[i\sum_{m=1}^{\infty}\frac{1}{\sqrt{m}}\,a_{m,\mu}z^{-m}\right]. \end{aligned} \tag{II.22}$$

Using eqs. (II.18), (II.19) one can derive the following operator equivalent of the KN-representation (II.16) (comp. fig. 3b)

$$\begin{aligned} B_N(1, 2, \ldots, N) = g^{-2} \lim_{\substack{z_1\to\infty\\ z_2\to 1\\ z_N\to 0}} \int\limits_{z_1 \leq z_2 \leq\ \leq z_N} \frac{1}{d^3V_{12N}}\left(\prod_{i=1}^{N} dz_i|z_i - z_{i+1}|^{a-1}\right) \\ \times\ \langle 0; 0|\left(\prod_{i=1}^{N} z_i^{-a}V_0(p_i, z_i)\right)|0; 0\rangle. \end{aligned} \tag{II.23}$$

---

[4]) To have a uniform notation, it is sometimes convenient to treat the momentum operator as a zero mode oscillator, i.e. $a_0{}^\mu = p^\mu$.

II.1.5. The projective properties of the vertex operator; The extension of the projective group

Both the Möbius invariance of the integrand and the related cyclic invariance of the representation (II.23) are out of question due to the built in identity to the KN-representation. Nevertheless we are going to discuss how these important Möbius invariance properties can be traced back to the behaviour of the vertex operator (II.21) under projective transformations.
The three-parameter projective group is given by the following algebra of generators ($L^+ = L_{-1}$, $L^- = L_{+1}$, L_0)

$$[L_m, L_n] = (m - n) L_{m+n} \qquad (m, n = \pm 1,0). \tag{II.24}$$

Obviously, L^+, L^- are 'raising' and 'lowering' operators with respect to eigenvalues of L_0, respectively. Any unitary representation of the group element γ, being isomorphic to the mapping $z' = \gamma(z)$, can therefore be written as

$$0(\gamma) = e^{i\xi L} = e^{i(\xi_+ L_1 + \xi_+^* L_{-1} + \xi_0 L_0)} \tag{II.25}$$

with ξL hermitean and ξ containing the three parameters of the 2×2 matrix $\hat{p}$ (eq. (II.14)). The representation of the projective algebra (II.24) in terms of oscillator operators had been found first by Gliozzi [22]. One introduces another Fubini-Veneziano field $P_\mu(z)$ by

$$P_\mu(z) = iz \frac{d}{dz} Q_\mu(z) = p_\mu + \sum_{n=1}^{\infty} \sqrt{n} \, (a_{r,\mu} z^{-n} + a_{n,\mu}^+ z^n). \tag{II.26}$$

The operatorial projective generators can then be written concisely as contour integrals

$$L_m = -\frac{1}{2} \oint \frac{dz}{2\pi i z} :P^2(z): z^m \equiv \left\langle -z^m \cdot \frac{1}{2} :P^2(z): \right\rangle \qquad (m = \pm 1,0) \tag{II.27}$$

where the contour encircles the origin and the normal product refers to the a_n-operators. Using eq. (II.27) one obtains, for example,

$$L_0 = -\frac{p^2}{2} - \sum_{n=1}^{\infty} n a_r^+ a_n = -\frac{p^2}{2} + H. \tag{II.28}$$

Operatorial functions of z (like $Q_\mu(z)$, $P_\mu(z)$, $V_0(p, z)$) can now be investigated with respect to their transformation behaviour under the projective group similarly as one classifies usual tensor fields according to their $O(3)$ spin. Let us consider an irreducible representation of the projective group characterized by the eigenvalue $J(J + 1)$ of the Casimir operator $L^2 = L_0(L_0 + 1) - L_1 L_{-1}$. Then, J is called its projective spin or projective weight. Furthermore, an operator function $X_j(z)$ is said to have projective spin J if it satisfies the following commutation relations with the generators (II.24)

$$[L_n, X_J(z)] = z^n \left[z \frac{d}{dz} - nJ \right] X_J(z). \tag{II.29}$$

It is then straightforward to prove that for a product of two operator functions the projective spin is additive

$$X_{J1} \cdot X_{J2} = (\overline{X})_{J1+J2}. \tag{II.30}$$

In particular, one easily verifies that the 'fields' $Q_\mu(z)$, $P_\mu(z)$ have projective spin $J = 0$ and $J = -1$, respectively. The vertex $V_0(p, z)$ of eq. (II.21) describing the absorption (or emission) of a scalar particle with mass m_0^2 turns out to have the projective spin $J_{V_0} = p^2/z = -a$ (we recall $p^2 = m_0^2 = -2a$). This statement is equivalent (eq. (II.29)) to the commutators

$$[L_n - L_0, V_0(p, z)] = \left\{(z^n - 1)\, z \frac{d}{dz} + z^n na\right\} V_0(p, z) \tag{II.31}$$

and to

$$V_0(p, z) = z^{L_0} V_0(p, 1)\, z^{-L_0}. \tag{II.32}$$

The latter equation is independent of the value of the projective spin and is related only to the case $n = 0$ in eq. (II.29). The equations (II.31) and (II.32) are the operatorial reflection of the projective properties of the integrand of the B_N-function in eq. (II.23) which all the essential properties of the GVM can be reduced to.
For the dual theory the discovery by VIRASORO [23] of a larger algebra containing the projective generators $L_{\pm 1,0}$ as a subalgebra was of great importance. The infinite set of 'Virasoro operators' L_m is given by eq. (II.27) generalized to an arbitrary integer index m. The commutator algebra of these L_m 's turns out to be

$$[L_m, L_n] = (m - n)\, L_{m+n} + \delta_{m,-n} \frac{D}{12}\, m(m^2 - 1). \tag{II.33}$$

Here D is the number of space-time dimensions which we abstain to fix to the physical value 4 for reasons that will become clear later. As a whole, the algebra (II.33) is related to all the conformal transformations in the z-plane and is therefore sometimes referred to as conformal algebra. One may now generalize the notion of projective spin to conformal spin. Eq. (II.29) defines a operator function of conformal spin J, if it is valid for any integer n and J independent of n. In particular, it turns out that the operator functions $Q_\mu(z)$, $P_\mu(z)$ and $V_0(p, z)$ all have a definite conformal spin, the values of which are known from the discussion of the projective properties ($J_Q = 0$, $J_P = -1$, $J_{V_0} = -a$). For the vertex operators that build up essentially the whole dual tree amplitudes there is now an infinity of commutation relations (eq. (II.31)) with arbitrary integer index n. Just these relations will later allow for an elimination of all the ghost states, if we put $a = 1$.
We would like to stress here that the duality properties are linked only with the proper Möbius group containing L_0, $L_{\pm 1}$ as generators. The larger conformal group is not an invariance group of the integrand of the KN-representation, as neither the vacuum state $|0; 0\rangle$ is invariant nor the integration measure behaves covariantly under the conformal group. Contrarily, the projective invariance of the vacuum can easily be shown by using the explicit expressions of $L_{\pm 1}$, L_0.

II.1.6. Factorization and the spectrum of states

For the discussion of factorization is it useful to remove the z-dependencies of the vertex operators by the relation (II.32) putting all $N - 3$ integrations in the real KN-representation (II.23) formally into $N - 3$ factorized propagator operators inserted between the vertices $V_0(p_i, 1)$. One obtains

$$B_N = \langle 0; -p_1|\, V_0(p_2, 1)\, D V_0(p_3, 1) \ldots D V_0(p_{N-1}, 1)\, |0; p_N\rangle \tag{II.34}$$

where D denotes the integrated propagator

$$D = \int_0^1 dx x^{-a-1+L_0}(1 - x)^{a-1} \equiv B(L_0 - a, a) = \sum_{l_0=0}^{\infty} \binom{l_0 - a}{l_0} \frac{1}{L_0 - a + l_0} \tag{II.35}$$

and we have introduced the states

$$|0; p\rangle = e^{-ipx}|0; 0\rangle. \tag{II.36}$$

Any manipulations with integrated operators are quite formal here as in practice all the operator algebra has to be performed before the integrations. Anyway, integrated operators are useful abbreviations providing a compact formalism. Using the commutation relation (II.19) one can further eliminate the zero mode operators p_μ and x_μ from eq. (II.34). This yields further (comp. fig. 3b)

$$B_N = \langle 0| V_0(p_2) D(s_{12}) V_0(p_3) D(s_{13}) \dots D(s_{1,N-2}) V_0(p_{N-1}) |0\rangle \tag{II.37}$$

where $|0\rangle$ denotes now the vacuum for the oscillator operators only and the propagator and vertex operators without zero mode operators read

$$D(s_{1j}) = \int_0^1 dx x^{-\alpha(s_{1j})-1}(1-x)^{a-1} x^H \tag{II.38}$$

$$V_0(p) = e^{-ipQ^+(1)}e^{-ipQ^-(1)} \equiv e^{-(a^+|p)}e^{(a|p)}. \tag{II.39}$$

In the vertex we have introduced for notational convenience infinite vectors with the components

$$\{(a^+|\}_n{}^\mu = \{|a^+)\}_n{}^\mu = a_n{}^{+\mu}$$

$$\{|p)\}_n{}^\mu = p^\mu\{|1)\}_n = p^\mu \cdot \frac{1}{\sqrt{n}}. \tag{II.40}$$

We remark that the case $a = 1$ obviously plays a special role as in that case the whole sum over l_0 in eq. (II.35) is reduced to the $l_0 = 0$ term. Then, the propagator with or without zero mode operators reads

$$D^{a=1} = \frac{1}{L_0 - 1}, \qquad D^{a=1}(s) = \frac{1}{H - \alpha(s)}. \tag{II.41}$$

It was apparent already in the structure of the KN-representation (II.16′) that only for $a \neq 1$ there is a special correlation between neighbouring KN-variables due to the factors $|z_i - z_{i+1}|^{a-1}$ while for $a = 1$ there is an enlarged symmetry of the integrand. We shall see soon that this enlarged symmetry activates in a well defined manner the whole Virasoro algebra. This is also the reason why we shall henceforth mainly discuss the case $a = 1$.

Originally, the factorization properties of the Veneziano model have been derived by starting directly from the Bardakci-Ruegg integral representation [*24*]. A more elegant way is provided by using the operator formalism. To show this, we write down the spectral decomposition of the propagator (at c-number momentum p)

$$\frac{1}{L_0 - 1} \to \sum_{l=0}^{\infty} \sum_{k=1}^{d(l)} \frac{|\{l\}; p\rangle_k \, {}_k\langle\{l\}; -p|}{l - \alpha(p^2)} \tag{II.42}$$

where the occupation number states $|\{l\}; p\rangle$

$$|\{l\}, p\rangle = \prod_{n=1}^{\infty} \prod_{\mu_n=0}^{D-1} \frac{1}{\sqrt{l_{n,\mu_n}!}} (a_n{}^{+\mu_n})^{l_{n,\mu_n}} e^{-ipx}|0; 0\rangle \tag{II.43}$$

(p here c-number, x operator!)

are eigenstates of H to the eigenvalue l

$$l = \sum_{n=1}^{\infty} n \sum_{\mu_n=0}^{D-1} l_{n,\mu_n}. \tag{II.44}$$

By the notation $|\{l\}; p\rangle_k$ $(k = 1, 2 \ldots d(l))$ in eq. (II.42) we have made explicit the degeneracy of the eigenstates of H. The degree of degeneracy $d(l)$ is just given by the number of ways one can represent l in the manner of eq. (II.44). Inserting eq. (II.42) between the vertices $V_0(p_i, 1)$ and $V_0(p_{i+1}, 1)$ yields the following representation of the N-point function (II.34) in terms of $(1, i)$-channel poles

$$B_N(1, 2, \ldots, N) = \sum_{l=0}^{\infty} \sum_{k=1}^{d(l)} \langle 0, -p_1| \, V_0(p_2, 1) \, D \ldots D V_0(p_i, 1) \, |\{l\}, p\rangle_k$$

$$\times \frac{1}{l - \alpha(p^2)} \langle\{l\}, -p| \, V_0(p_{i+1}, 1) \, D \ldots D V_0(p_{N-1}, 1) \, |0; p_N\rangle \tag{II.45}$$

$$\left(p = \sum_{j=1}^{i} p_j = -\sum_{j=i+1}^{N} p_j; \qquad p^2 = s_{1i}\right).$$

At each level with $\alpha(s_{1i}) = l$ one finds a sum of factorizing residua, the number of which reflects the degeneracy of the spectrum. Again the level number l gives the maximal spin j of the state on the leading or 'parent' (P) trajectory $j = \alpha(p^2)$. The states lying on this P-trajectory turn out to be nondegenerate [8] and are totally made up by excitations of the space-like a_1-modes ($\boldsymbol{a}_1$)

$$j^{(P)} = l = \sum_{\mu_1=1}^{D-1} l_{1,\mu_1}. \tag{II.46}$$

For $a > 0$, especially for the Virasoro case $a = 1$ unfortunately there is a scalar tachyon state ($m^2 < 0$) lying on the parent trajectory (see eq. (II.3)) Furthermore, due to the degeneracy of the spectrum there is an infinity of daughter trajectories $\alpha_d(p^2) = \alpha(p^2) - d(d = 1, 2, \ldots)$ as can be verified by 0(3) reduction.

By eq. (II.45) the GVM reveals a remarkable degree of selfconsistency. Each of the factors in a residuum term at a pole $\alpha(p^2) = l$ is itself again of the structure of a Veneziano amplitude. For $l > 0$ it has one external excited state $|\{l\}, p\rangle_k$ containing spins $j \leqq l$.

In fig. 4 we illustrate the factorization in terms of quarkline diagrams (comp. fig. 2).

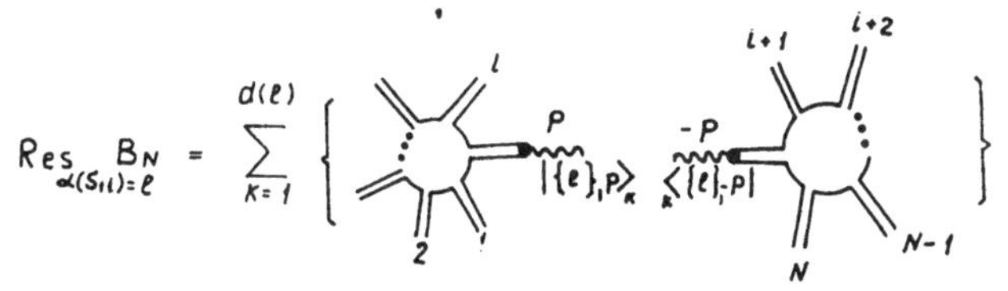

Fig. 4. The factorization of the N-point function in terms of quark-line diagrams

The factorization can be continued to produce tree amplitudes with an arbitrary number of external excited particles. One obtains this way operatorial expressions for the three — and generally N-Reggeon vertices (see chapt. II.1.9). These elements can then be used as building blocks for writing down an arbitrary diagram, in principle. Especially, they are the necessary starting points for constructing unitarity corrections via loop diagrams (comp. chapt. III).

Concluding this section, we mention the fact that one obtains from the level degeneracy $d(l)$ the following asymptotic exponential behaviour of the number $\tilde{d}(m)$ of states per unit mass (mass $m \sim \sqrt{l}$) [24]

$$\tilde{d}(m) \sim m^{-D+1/2} \exp\left(m\pi \sqrt{\frac{2\alpha' D}{3}}\right). \tag{II.47}$$

The behaviour (II.47) coincides qualitatively with that obtained by HAGEDORN [25] on the basis of the statistical bootstrap model. This fact is a remarkable coincidence of results based on rather different physical concepts. It has been the starting point for interesting investigations of the link between the dual resonance and the thermodynamic model [26].

II.1.7. Gauge conditions and no-ghost theorem

Up to now the outcome of the factorization of dual tree amplitudes seems to be that all the occupation number states $|\{l\}, p\rangle$ in eq. (II.45) are present as intermediate states and contribute to the residua. A thorough investigation exhibits, however, that this is definitely not the case. Indeed, there are states $|s\rangle$ in the Fock space of the harmonic oscillators which do not couple at the level $l = \alpha(p^2)$ to the 'tree'-states

$$|t; p\ , p_2, \ldots, p_N\rangle = V_0(p_1, 1)\, D V_0(p_2, 1) \ldots V_0(p_{N-1}, 1)\, |0; p_N\rangle \tag{II.48}$$

being therefore spurious, i.e.

$$\langle s|\, t; p_1, p_2, \ldots, p_N\rangle = 0. \tag{II.49}$$

Expanding the spurious states in the occupation number basis, we obtain from eq. (II.49) linear dependencies between matrix elements $\langle s|\, \{l\}, p\rangle$ with different excited states $|\{l\}, p\rangle$. These relations are analogous to the Ward identities reflecting subsidiary relations for physical states in a gauge field theory. Let us now briefly demonstrate how one obtains in the Virasoro case $a = 1$ an infinity of subsidiary gauge conditions for the 'physical states' producing nonzero contributions at the pole $l = \alpha(p^2)$ to the tree state (II.48). We quote first some useful commutation relations. By the Virasoro algebra (II.33) one gets (recall eq. II.41)

$$(L_n - L_0 + 1) \frac{1}{L_0 - 1} = \frac{1}{L_0 + n - 1} (L_n - L_0 + 1 - n) \tag{II.50}$$

and, using eqs. (II.31), (II.50), also

$$(L_n - L_0 + 1) \frac{1}{L_0 - 1} V_0(p, 1) = \frac{1}{L_0 + n - 1} V_0(p, 1)\, (L_n - L_0 + 1) \tag{II.51}$$

so that

$$\begin{aligned}(L_n - L_0 + 1) &\frac{1}{L_0 - 1} V_0(p_1, 1) \frac{1}{L_0 - 1} \cdots \frac{1}{L_0 - 1} V_0(p_{N-1})\, |0; p_N\rangle \\ &= \frac{1}{L_0 + n - 1} V_0(p_1, 1) \frac{1}{L_0 + n - 1} \cdots \frac{1}{L_0 + n - 1} \\ &\quad \times V_0(p_{N-1}, 1)\, (L_n - L_0 + 1)\, |0; p_N\rangle = 0.\end{aligned} \tag{II.52}$$

In the last equation we have used the mass shell condition of the ground state $(L_0 - 1)\, |0; p_N\rangle = 0$ as well as the fact $L_n |0; p_N\rangle = 0$. Eq. (II.52) means that a physical on-shell state $|\Phi_{0n}\rangle$ which couples at the pole $l = \alpha(p^2)$ to the tree state (II.48) ($\langle \Phi_{0n}|\, \text{tree}\rangle \neq 0$)

satisfies

$$(L_n - L_0 + 1)\,|\Phi_{0n}\rangle = 0. \tag{II.53}$$

Using the mass-shell condition $D^{-1}\,|\Phi_{0r}\rangle = (L_0 - 1)\,|\Phi_{0n}\rangle = 0$ one concludes further that an arbitrary physical state necessarily statisfies the conditions

$$(L_0 - 1)\,|\Phi_{0n}\rangle = 0 \tag{II.54a}$$

$$L_n|\Phi_{0n}\rangle = 0 \qquad (n = 1, 2, \ldots). \tag{II.54b}$$

Due to eq. (II.50) one can write eq. (II.52) also in the form

$$W_n|\text{tree}; p\rangle \equiv (L_n - L_0 + 1 - n)\,|\text{tree}; p\rangle = 0. \tag{II.55}$$

The spurious states decoupling from any tree are therefore given by

$$\langle s_n| = \langle\psi|\, W_n \qquad (n = 1, 2, \ldots) \tag{II.56}$$

($\langle\psi|$ arbitrary). They are orthogonal to the solutions of eq. (II.54) as can easily be seen by introducing the projector onto the lth level

$$\mathfrak{M} = \oint \frac{dz}{2\pi i z}\, z^{H-l} \tag{II.57}$$

and writing

$$\langle s_n|\Phi_{0n}\rangle = \langle\psi|\, W_n\mathfrak{M}_l\,|\Phi_{0n}\rangle = \langle\psi|\,\mathfrak{M}_{l-n}L_n\,|\Phi_{0n}\rangle = 0. \tag{II.58}$$

The gauge equation (II.54b) are very important because they are the key for solving the 'ghost' difficulties of the dual resonance model. These ghosts are negative norm states which appear through the timelike oscillators, e.g. $a^+_{n,0}\,|0\rangle$, because of the manifestly Lorentz-covariant algebra (II.18). A similar situation is well known from the covariant formulation of quantum electrodynamics where Lorentz-covariance can be reconciled with positivity of the Fock space by imposing the Lorentz condition as a 'weak' subsidiary condition on the space of physical states [27]. There, the time-like photons combine with longitudinal photons to decoupling spurious or null norm states leaving us with the transverse photons as the only effectively coupling physical states. Now, in the case of the DRM an infinity of gauge relations is needed in order to eliminate the ghosts of the infinite set of timelike oscillators $a^+_{n,0}$. As has been first suggested by VIRASORO [23] this infinity of gauge relations is given by eq. (II.54b). The proof that the solutions of eq. (II.54a, b) have indeed nonnegative norm is highly nontrivial. It is the subject of the so-called no-ghost theorem being one of the major outcomes of the algebraic approach to the dual model.

As there is an exhaustive literature on the proof of the no-ghost theorems [12, 28] we shall restrict ourselves to a short presentation of the substantial ideas involved. An important first step was done by DEL GIUDICE, DI VECCIA and FUBINI (DDF) [28] who constructed explicit solutions to eq. (II.54a, b) by using a non-covariant method analogous to the Coulomb gauge in QED. In particular, these authors introduced 'transverse' operators

$$A_n{}^i = \oint \frac{dz}{2\pi i z}\, \varepsilon^i P(z)\, e^{-inkQ(z)} \begin{pmatrix} i = 1, 2, \ldots, D-2 \\ n \gtreqless 0 \end{pmatrix} \tag{II.59}$$

where the vectors $\varepsilon^i (i = 1, \ldots, D - 2)$ are transverse polarization vectors with respect to the direction of k, i.e. $\varepsilon^i \cdot k = 0$, and $k^2 = 0$. Then, the following relations can easily be derived

$$[A_m{}^i, A_n{}^j] = m\delta_{ij}\delta_{m,-n} \tag{II.60}$$

$$[L_m, A_n{}^i] = 0. \tag{II.61}$$

The so-called DDF-states defined in analogy to eq. (II.43) by ($p_0k = 1$)

$$|DDF, \{\lambda\}\rangle = \prod_{n=1}^{\infty} \prod_{\mu_n=1}^{D-2} \frac{1}{\sqrt{\lambda_{n,\mu_n}!}} \left(\frac{1}{\sqrt{n}} A_{-n}^{\mu_n}\right)^{\lambda_{n,\mu_n}} |0; p_0\rangle \tag{II.62}$$

are in fact orthonormalized physical on-shell states. The orthogonality and proper normalization follows from the isomorphism between the DDF-algebra (II.60) and the algebra of the 'transverse' original oscillators $a_n{}^i$. The property of the state to be physical is guaranteed by eq. (II.61) leading to L_n |DDF⟩ = 0. Due to eq. (II.60) the DDF subspace has a positive definite metric. Obviously the DDF-operators are at least part of the spectrum generating algebra (SGA), defined to generate all solutions of eq. (II.54a, b) out of the ground state. BROWER [28] completed the SGA by adding two sets $A_n{}^+$, $A_n{}^-$ of destruction ($n > 0$) and creation ($n < 0$) operators. They take account of the timelike and longitudinal directions of k_μ and are analogously constructed as the $A_n{}^i$ in order to commute with all the L_n's. The essential points now are that one set becomes trivial ($A_r{}^+ = \delta_{n,0}$) while the other ($A_n{}^-$) can be seen to create physical states that are contained in the DDF-states of a Minkowski space with a 'critical' dimensionality $D_c = 26$. They are, therefore, positive norm states, too. As long as $D < 26$, the $A_n{}^-$-created physical states are not purely transverse DDF states in the D dimensional space-time, but contain longitudinal modes. For the 'critical' dimension $D = D_c = 26$, however, the norm of the latter ones becomes zero and these zero norm physical states decouple similar as the null norm states of the covariantly quantized (four-dimensional) electrodynamics. For $D > 26$ one finds immediately ghosts created by the $A_n{}^-$'s. In this sense, only the case $D = D_c$ corresponds to the situation in QED with both scalar and longitudinal photons decoupling.

The proof of the no-ghost theorem for $D \leqq 26$ was completed by GODDARD and THORN [28] by analysing the whole Fock space of DDF-states, of negative norm spurious states and of the null norm sector containing both spurious and physical parts. It could be shown that for the critical dimensionality $D = 26$ any physical state contributing to a residuum is represented by a DDF-state and a decoupling state of zero norm. This way every physical state at $D = 26$ is equivalent to a DDF-state.

We would like to emphasize here that the equation (II.31) together with the condition $a = 1 (J_{V_0} = -1)$ was decisive for obtaining the infinity of gauge conditions (II.54b)[5]. The price one has to pay for getting rid of the ghost states ($D \leqq 26$) consists therefore in the introduction of an unphysical tachyon ground state. It is the optimistic point of view of many people that it will be easier to remove such a tachyon one day than to eliminate ghosts in a totally different model. In this sense the 'automatic' ghost killing mechanism of the dual theory should be considered as one of its major virtues. The tachyon elimination may well take place along the lines of spontaneous symmetry breaking which should lead as usual to a shift in the ground state mass [29].

II.1.8. The twist operator

For completeness and for further use in later sections we are going to discuss in this paragraph the so-called twist operation, represented by an operator Ω which is defined to

[5]) It has been shown [8] that for $a \neq 1$ only the first (L_1) gauge condition is effective rendering only the first daughter trajectory ghost-free.

perform in a tree state (II.48) the following cyclic change of the external particles

$$\Omega\,|t; p_1, p_2, \ldots, p_{N-1}, p_N\rangle = |t; p_N, p_{N-1}, \ldots, p_2, p_1\rangle \qquad \text{(II.63)}$$

Fig. 5 is to illustrate that Ω performs just a 'twist' (symbolized by a cross) of the internal line it acts on.
An explicit form of Ω can easily be obtained by performing a projective transformation in the integrand of the tree state configuration of the right-hand side of eq. (II.63). The operator Ω is essentially the representation $0(\gamma(z))$ (eq. II(25)) of that transformation. It reads [30]

$$\Omega = (-1)^H\, e^{-L_{-1}} = e^{L_{-1}}(-1)^H. \qquad \text{(II.64)}$$

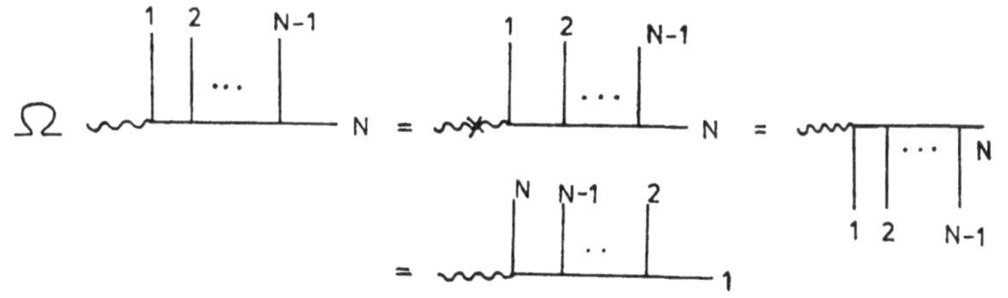

Fig. 5. Graphical representation of the twist operation

It seems plausible to demand that a doubly twisted line with a propagator should be identical to the propagator alone. Unfortunately the propagator D is not yet double twist invariant, i.e.

$$\Omega^+ D\Omega \neq D \qquad \text{(II.65)}$$

due to the presence of spurious states. Because of eq. (II.65) the twisted propagator is then also not hermitean, $(D\Omega)^+ \neq D\Omega$. A hermitean twisted propagator D_T can however be defined by including a 'gauge transformation' $(1-x)^{(L_0-L_1)} = (1-x)^{-w_1}$ [8]

$$D_T = \int_0^1 dx\, x^{-a-1+L_0}(1-x)^{a-1}\,\Omega(1-x)^{-w_1}. \qquad \text{(II.66)}$$

II.1.9. Canonical forms; N-Reggeon vertex

We would like to add some remarks on the general structure of operators involved in building up tree amplitudes that will again shed some light on the special role played by the projective group. It is useful to define first a canonical form C as having the following structure of a normal ordered product of exponential operators [8].

$$C = e^{-(a^+|A)} : e^{-(a^+|X-1|a)} : e^{-(a|B)}\, e^{-\varphi}. \qquad \text{(II.67)}$$

Here $|A)$, $|B)$ and X are definite vectors and an infinite quadratic matrix in the notation of eq. II.40, respectively, φ is a c-number. It turns out that for all operators of the form of eq. (II.67) being relevant in the dual model the infinite matrices X can be defined in terms of projective transformations $\gamma(z)$ (Comp. eq. (II.13)), i.e.

$$[X\{z\}\,|1)]_m \equiv \sum_{n=1}^{\infty} X_{mn} z^n \frac{1}{\sqrt{n}} = \big[[\{\gamma(z)\} - \{\gamma(0)\}]\,|1)\big]_m = \frac{1}{\sqrt{m}}\big(\gamma^m(z) - \gamma^m(0)\big) \qquad \text{(II.68)}$$

where we have used the notation $\{z\}_{mn} = \delta_{mn} z^m$. Similar representations can be found for the other elements of the canonical form C. In this way, a canonical form may be characterized by its respective 2×2 matrix $\hat{p}$ or, equivalently, by the Lovelace symbol (II.15). For example, the integrands of the untwisted and twisted operators D, $D_{T'}$ may be written as canonical forms (II.67) with the following attached projective transformations [19]

$$x^{L_0} \triangleq \begin{bmatrix} 0 & \infty & 1 \\ 0 & \infty & x \end{bmatrix} \tag{II.69}$$

$$x^{L_0}\Omega(1-x)^{-w_1} \triangleq \begin{bmatrix} 0 & \infty & 1 \\ x & 1 & 0 \end{bmatrix}. \tag{II.70}$$

For later use we quote also the explicit form of the symmetrical N-Reggeon vertex operator V_N which can be obtained by factorizing tree amplitudes with a least $2N$ external particles. (see fig. 6. The dots on the lines indicate the orientation of the attached trees.) The vertex operator V_N as given by OLIVE [31] can then be written formally as a $2N$ dimensional KN-integral with two variables z_i, β_i belonging to each external leg. The usual measure d^3V_{abc} for three

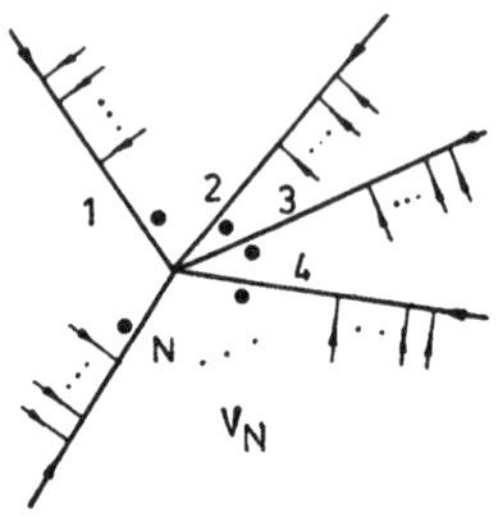

Fig. 6. Possible factorization leading to the N-Reggeon vertex

fixed z-variables is divided out and the $2N$ KN-variables are ordered according to $z_1 \to \beta_1 \to z_2 \to \beta_2 \to \cdots \to z_N \to \beta_N \to z_1$ on the KN-circle. To each leg a real (Chantype) variable is related, defined by (comp. eq. (II.11)) the anharmonic ratio

$$x_i = (\beta_i, z_{i+1}, z_i, \beta_{i-1}) \tag{II.71}$$

which is fixed, i.e. N integrations are undone again by dividing out formally all dx_i. The net result is as follows

$$V_N = g \oint\limits_{z_1 \to \beta_1 \to \cdots \to z_1} \frac{1}{d^3V_{abc}} \left\{ \prod_{i=1}^{N} dz_i \, d\beta_i \left(\frac{z_{i+2}-\beta_i}{z_{i+1}-\beta_i}\right)^{a-1} \left(\frac{1-x_i}{dx_i}\right) \right\}$$
$$\times \exp\left\{ -\frac{1}{2} \sum_{\substack{i,j=1 \\ (i \neq j)}}^{N} (a^{(i)+} | U^i V^j | a^{(j)+}) \right\} \tag{II.72}$$

where the U^i, V^j are infinite dimensional representations of special projective transformations in the sense of eq. (II.68)[6]).

[6]) Eq. (II.72) includes the zero modes $a_0^{(i)+} = p^i$. Therefore, the matrices used here are defined to include the rows and columns U_{0n}, U_{n0} $(n > 0)$ and the element U_{00} which may also be expressed by the same projective transformation $\gamma(z)$ as the elements U_{mn}.

They are defined as

$$U^i \triangleq \hat{U}^i = \begin{pmatrix} \zeta_i & -\zeta_i\beta_{i-1} \\ 1 & -z_i \end{pmatrix} \triangleq \begin{bmatrix} \beta_{i-1} & z_i & z_{i+1} \\ 0 & \infty & 1 \end{bmatrix}$$

$$V^j \triangleq \hat{V}^j = \begin{pmatrix} \zeta_j\beta_{j-1} & -z_j \\ \zeta_j & -1 \end{pmatrix} \triangleq \begin{bmatrix} \infty & 0 & 1 \\ \beta_{j-1} & z_j & z_{j+1} \end{bmatrix} \tag{II.73}$$

$$\left(\zeta_i = \frac{z_i - z_{i+1}}{\beta_{i-1} - z_{i+1}}\right).$$

II.1.10. Introduction of internal quantum numbers by the Chan-Paton method

Chan and Paton [32] proposed an effective and economic way of introducing quantum numbers into the dual theory. Their recipe takes into account both the factorization properties and the absence of exotic states so that some underlying quark-antiquark structure for boson resonances is implied. Let us assume the group $SU(N)$ as the internal symmetry group (e.g. $SU(2)$ for isospin or $SU(3)$ for the conventional quark model). If the external particles of an individual planar Veneziano term are assumed to belong to the $(N^2 - 1)$-plet of the regular representation or to the singlet, they are represented by one of the N^2 λ_i-matrices $\left(i = 1, 2 \ldots N^2 - 1,\ \lambda_0 = 1 \cdot \sqrt{2/N}\right)$ corresponding to the generators and the unit matrix in the regular representation. They are related to the Clebsch-Gordan coefficients $\begin{pmatrix} N & N & N^2 - 1 \\ \nu' & \nu & i \end{pmatrix} \equiv (\lambda_i)_{\nu'\nu}$ for the coupling to a q$\bar{\text{q}}$-pair. These matrices are normalized by $1/2 \operatorname{Tr}(\lambda_i\lambda_j) = \delta_{ij}$ and satisfy the completeness relation

$$\frac{1}{2} \sum_{i=0}^{N^2-1} (\lambda_i)_{\nu'\nu} (\lambda_i)_{\mu'\mu} = \delta_{\nu'\mu}\delta_{\nu\mu'}. \tag{II.74}$$

Now the rule for a single planar Veneziano term with a given ordering $p_1, p_2, \ldots, p_N$ of the external particles consists in multiplying $B_N(1, \ldots, N)$ by the trace

$$\frac{1}{2} \operatorname{Tr}(\lambda_1 \cdot \lambda_2, \ldots, \lambda_N) \tag{II.75}$$

preserving autmatically the cyclic symmetry of $1, 2, \ldots, N$. By eq. (II.74) one can factorize the considered term according to

$$\frac{1}{2} \operatorname{Tr}(\lambda_1 \cdot \lambda_2 \ldots \lambda_N) = \sum_{i=0}^{N^2-1} \left[\frac{1}{2} \operatorname{Tr}(\lambda_1\lambda_2 \ldots \lambda_k\lambda_i)\right] \left[\frac{1}{2} \operatorname{Tr}(\lambda_i\lambda_{k+1} \ldots \lambda_N)\right]. \tag{II.76}$$

This is consistent with the factorization of the momentum dependent part of B_N illustrated in fig. 4. All intermediate states are automatically combined with the same λ-matrices appearing already for the external particles excluding thus exotic intermediate states. As each intermediate state contains both singlet and $(N^2 - 1)$-plet contributions we have generally at least a twofold exchange degeneracy of each trajectory. Some applications of this technique can be found, for example, in ref. [33].

II.2. The Neveu-Schwarz-Ramond (NSR) model

II.2.1. Definition of the Neveu-Schwarz (NS-)pion model

In 1971 Neveu and Schwarz [3] discovered their dual 'pion' model which exhibits a remarkable enrichment of the spectrum relative to the GVM. In particular, there are now different families of trajectories ('ρ' and 'π, ω') in two and three particle channels with a spacing of one half between the π- and ρ-trajectory. The NS-spectrum shows an astonishing amount of similarity to the set of nonstrange resonances found in nature, though there remain yet some major defects. Thus, the π- and the ρ-trajectories come out half a unit too high, making the 'pion' a tachyon and the ρ massless. The π- and ω-trajectories are exchange degenerate. All ghosts are removed similarly as in the GVM case, but the critical dimensionality below which this happens with the spectrum being purely 'transverse' (DDF) is lowered to $D_c = 10$. An optimist is thus led to believe that we are on the right track towards $D_c = 4$ in a future model.

The enrichement of the spectrum is achieved by the introduction of an additional set b_k^μ, $b_k^{+\mu} = b_{-k}^\mu$ of 'oscillator' operators with half odd integer indices k which will be used in addition to the a_n^μ-operators as well as the position and momentum operators x^μ, p^μ known already from the Veneziano model. The b- and a-operators are defined to commute with each other, while the b-operators themselves satisfy an anticommutator algebra, so that they are in fact Fermi operators.

$$\{b_k^\mu, b_l^\nu\} = \{b_k^{+\mu}, b_l^{+\nu}\} = 0; \qquad \{b_k^\mu, b_l^{+\nu}\} = -g^{\mu\nu}\delta_{kl}$$
$$\left(k, l = \frac{1}{2}, \frac{3}{2}, \frac{5}{2}, \ldots\right). \tag{II.77}$$

The vacuum of the model is the zero eigenstate of all b_k^μ, a_m^ν $(k, m > 0)$

$$a_m^\nu |0\rangle = b_k^\mu |0\rangle = 0 \qquad (k = 1/2, 3/2, \ldots; \quad m = 1, 2, \ldots). \tag{II.78}$$

In analogy to the Fubini-Veneziano fields $Q_\mu(z)$, $P_\mu(z)$ one introduces the b-operator function $H_\mu(z)$ as a power series, but now with half odd integer powers of z:

$$H_\mu(z) = \sum_{k=1/2}^{\infty} (b_{k,\mu} z^{-k} + b_{k,\mu}^+ z^k) = \sum_{k=-\infty}^{+\infty} b_{k,\mu} z^{-n}. \tag{II.79}$$

The vertex operator for the absorption (or emission) of a scalar particle with momentum p_i is now defined to be

$$V^{(\mathrm{NS})}(p_i, z) = g_{\mathrm{NS}} p_i \cdot H(z) : e^{-ip_i Q(z)} : \tag{II.80}$$

where g_{NS} is a universal coupling constant. Then, one can write down an operator representation for the amplitude of the multiperipheral configuration of fig. 3b quite analogous to eq. (II.23) where only $V_0(p, z)$ is replaced by $V^{(\mathrm{NS})}(p, z)$ and $a = 1$. The duality properties of the B_N-function are again related to the projective invariance of the corresponding integrand. As in the GVM there are projective generators L_0, $L_\pm$ that are part of the larger set of 'conformal' generators $L_n (n \gtreqless 0)$. They are now a sum of the usual a-type operators L_n^a as defined in eq. (II.27) and new b-type operators L_n^b

$$L_n^{\mathrm{NS}} = L_n^a + L_n^b$$
$$L_n^b = \frac{1}{2} \oint \frac{dz}{2\pi i z} z^n : H(z) \cdot z \frac{d}{dz} H(z) : \qquad (n = 0, \pm 1, \ldots). \tag{II.81}$$

They satisfy a Virasoro algebra (comp. eq. II.33)

$$[L_m^{\mathrm{NS}}, L_n^{\mathrm{NS}}] = (m - n) L_{m+n}^{\mathrm{NS}} + \delta_{m,-n} \frac{D}{8} m(m^2 - 1) \tag{II.82}$$

with a modified c-number term. This change in the anomalous term will be responsible for lowering the critical dimension from 26 to 10 as has been found in deriving the no-ghost theorems of the NS-model [9]. Note that the field $H_\mu(z)$ is constructed to have conformal spin $J_H = -1/2$ under this algebra whereas the a-part of the vertex $:e^{-ipQ(z)}:$ has as in the GVM $J_{\Gamma_0} = p^2/2$. As the conformal spin is additive for a product (comp. eq. II.30) and the conformal spin of the whole vertex has to be -1 in order to get the L_n^{NS} $(n > 0)$ as gauge operators this enforces the following condition for the mass squared of the absorbed (emitted) external particle:

$$\frac{m^2}{2} = -\frac{1}{2}. \tag{II.83}$$

This state was not contained in the GVM. It defines the 'pion' of the model, being obviously a tachyon. Using a relation formally identical to eq. (II.32) one again puts all the $N - 3$ integrations in the operator representation of B_N into integrated propagators of the form (II.35) $(a = 1)$. This way one obtains the amplitude B_N in the so-called '$\mathfrak{F}_1$-representation'.

$$B_N^{(NS)}(1, \ldots, N) = \langle 0; -p_1| (p_1 \cdot b_{1/2}) \, V^{NS}(p_2, 1) \frac{1}{L_0^{NS} - 1} \cdots V^{NS}(p_{N-1}, 1) \, (p_N b_{1/2}^+) \, |0; p_N\rangle. \tag{II.84}$$

The $b_{1/2}$-operators at the 'boundary' states $\langle 0; -p_1|$, $|0; p_N\rangle$ are remnants of the factors $p_{1(N)} H(z)$ in the boundary vertices. The factorization properties of the amplitude $B_N^{(NS)}$ can be made explicit as in the GVM in a straightforward way. Furthermore, the physical on-shell states $|\Phi_{on}^{NS,1}\rangle$ satisfy in the $\mathfrak{F}_1$-picture equations analogous to eq. (II.54a, b) where, of course, L_n^{NS} instead of L_n is understood. We have, however, now a new infinity of ghosts introduced by the $b_{k,0}^+$ operators. Thus, a further set of subsidiary conditions and corresponding gauge operators is needed. As anticipated already bei Neveu and Schwarz, these new operators are bilinear forms in a- and b-operators given by

$$G_k = \oint \frac{dz}{2\pi i z} \, z^k P(z) \cdot H(z) \qquad (k = \pm 1/2, \pm 3/2, \ldots). \tag{II.85}$$

II.2.2. The $\mathfrak{F}_2$-reformulation of tho NS-model, the spectrum

The spectrum of the amplitude in the $\mathfrak{F}_1$-representation (II.84) contains formally the usual GMV (tachyon) ground state $|0; \tilde{p}\rangle$ because of $(L_0^{NS} - 1)\,|0; \tilde{p}\rangle = 0$ $\left(\alpha(\tilde{p}^2) = 1 + 1/2\,\tilde{p}^2 = 0\right)$. It can be shown that this state decouples from the amplitude. To make this explicit, it is convenient to transform to another Fock space '$\mathfrak{F}_2$' with the 'pion' as ground state.

The reformulation is based on the observation that the operators $(p_2 \cdot b_{1/2})$ and $(p_N \cdot b_{-1/2})$ in eq. (II.84) can be replaced identically by $G_{1/2}$ and $G_{-1/2}$, respectively [3]. These G-operators then can be pushed to the opposite end of the operator product where they are annihilated by the vacuum due to $G_k|0; p\rangle = 0$ $(k \geq 1/2)$. The following commutators and anticommutators have to be used

$$\left.\begin{aligned} [L_m^{NS}, G_k] &= \left(\frac{m}{2} - k\right) G_{m+k} \\ \{G_k, G_l\} &= 2L_{k+l}^{NS} + \frac{D}{2}\left(k^2 - \frac{1}{4}\right)\delta_{k,-l} \end{aligned}\right\} \begin{aligned} &k, l = \pm 1/2, \pm 3/2, \ldots \\ &m = 0, \pm 1, \pm 2, \ldots \end{aligned} \tag{II.86}$$

The result reads as follows

$$B_N^{(\mathrm{NS})}(1, \ldots, N) = \langle 0; -p_1 | V^{\mathrm{NS}}(p_2, 1) \frac{1}{L_0^{\mathrm{NS}} - \frac{1}{2}}$$

$$\times V^{\mathrm{NS}}(p_3, 1) \ldots \frac{1}{L_0^{\mathrm{NS}} - \frac{1}{2}} V^{\mathrm{NS}}(p_{N-1}, 1) |0; p_N\rangle. \quad \text{(II.87)}$$

In deriving this representation use has been made of the fact that any term arising in the course of calculation with a propagator cancelled can be left aside as it gives a zero contribution due to the meromorphy of the amplitude in all channels. Eq. (II.87) shows explicitely the appearance of a new propagator, whereas the vertices remain unchanged in both pictures. It is now straightforward to prove that any physical on-shell state in the $\mathfrak{F}_2$-picture has to fulfil the constraints

$$\left(L_0^{\mathrm{NS}} - \frac{1}{2}\right) |\Phi_{\mathrm{on}}^{\mathrm{NS},2}\rangle = 0$$

$$L_n^{\mathrm{NS}} |\Phi_{\mathrm{on}}^{\mathrm{NS},2}\rangle = G_k |\Phi_{\mathrm{on}}^{\mathrm{NS},2}\rangle = 0 \qquad (n = 1, 2, \ldots, k = 1/2, 3/2, \ldots). \quad \text{(II.88)}$$

We mention that there is also some conserved G-parity present in the NS-model which is related to the number of b-modes contained in a state. In the $\mathfrak{F}_2$-picture the G-parity operator reads

$$G = -(-1)^{\sum_{k=1/2}^{\infty} b_k^+ b_k}. \quad \text{(II.89)}$$

The presence of a G-parity combined with the introduction of isospin by the Chan-Paton procedure allows the association of quantum numbers to the NS-states that can be tentatively identified with those of corresponding physical particle states. In fig. 7 we give for illustration the result of this identification for the lowest lying NS-states.

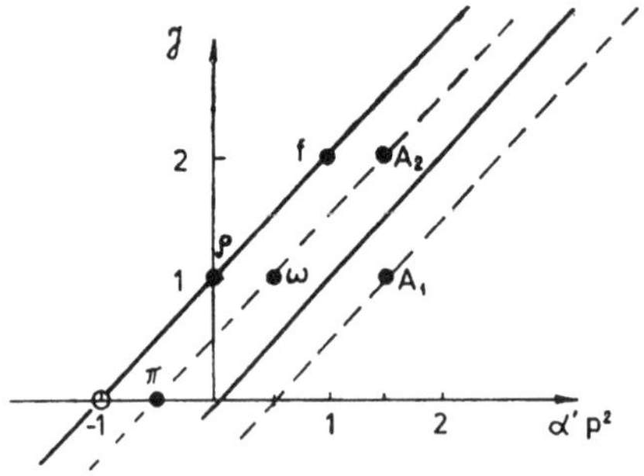

Fig. 7. Chew-Frautschi plot of the spectrum of the NS-model

There are in the lower part of the spectrum only those particles found in nature though these sometimes appear at wrong mass values. In sect. II.2.6. we shall calculate some amplitudes with external pions. There it will become obvious that

$$\alpha_\varrho(t) = 1 + \frac{1}{2} t, \qquad \alpha_{\pi,\omega}(t) = \frac{1}{2} + \frac{1}{2} t.$$

As a rule only the trajectories containing normal parity couplings $P = G(-1)^j$ like π, ρ, f are lying half a unit too high, while that part of the spectrum with abnormal parity coupling $P = -G(-1)^j$ (like ω) is approximately at the right position.
The four pion amplitude of the NS-Model has been shown to coincide with the phenomenological Lovelace-Shapiro pion amplitude [34] exposing formally the Adler zero for $\alpha(t) = \alpha_\rho(t) \xrightarrow[(t\to m_\pi^2)]{} 1/2$ and $\alpha_\rho(s) \xrightarrow[(s\to m_\pi^2)]{} 1/2$ due to the right spacing of half a unit between α_ρ and α_π. In particular, the $\pi\pi\pi\omega$ amplitude is exactly that one proposed by Veneziano in his initial paper [1].

II.2.3. The Ramond fermions and their interaction with Neveu-Schwarz bosons

In finding a suitable way of introducing fermions into the dual theory Ramond [4] was guided by some kind of correspondence principle that becomes visible already at the level of the GVM with $a = 1$. This principle leads essentially to the following replacement of the free Dirac equation

$$[\gamma p - m_F]\,\psi(p) = 0 \Rightarrow [\langle \Gamma(z) \cdot P(z)\rangle - m_F]\,|\psi_{\text{on}}^{R,2}\rangle = 0 \tag{II.90}$$

where $\Gamma_\mu(z)$ is a new 'field' generalizing the γ_μ-matrix, $\langle\cdots\rangle$ is the averaging procedure explained in eq. (II.27) and the index R, 2 stands for Ramond fermion state in the $\mathfrak{F}$-formulation which will be explained in the following. The field $\Gamma_\mu(z)$ turns out to be given again by a power series ($\gamma_5 = \gamma_0\gamma_1 \cdot \gamma_2 \cdots \gamma_{D-1}$; $\gamma_5{}^2 = 1$)

$$\Gamma_\mu(z) = \gamma_\mu + i\sqrt{2}\,\gamma_5 \sum_{n=1}^{\infty} (d_{n\mu} z^{-n} + d_{n\mu}^+ z^n) \tag{II.91}$$

where the algebra of the d-operators is formally identical to that of the b-modes in the NS-Model but the indices are integers here and there is *a* 'zero mode' $d_{0\mu} = 1/(i\sqrt{2})\,\gamma_\mu$

$$\{d_{m\mu}, d_{n\nu}^+\} = -g_{\mu\nu}\delta_{m,-n} \qquad (m, n = 0, 1, 2, \ldots). \tag{II.92}$$

Note that there is a formal correspondence (comp. eq. (II.79))

$$\Gamma_\mu(z) \leftrightarrow i\sqrt{2}\,H_\mu(z) \tag{II.93}$$

with the b-operator function $H_\mu(z)$ of the NS-model which is helpful in deriving analogous quantities; but care must always be taken with the zero mode that is present in the Ramond case only. The objects exactly corresponding to the Operators $(-G_k)$ in the sense of eq. (II.93) are the mixed a, d operators

$$F_n = -\frac{1}{i\sqrt{2}}\,\langle z^n \Gamma(z) \cdot P(z)\rangle. \tag{II.94}$$

(The d-operators are defined to commute with all the a-operators.) The spectral condition of 'Ramond-Dirac' equation (II.90) for the on-shell Ramond states now becomes

$$\left(F_0 - \frac{i m_F}{\sqrt{2}}\right) |\psi_{\text{on}}^{R,2}\rangle = 0. \tag{II.95}$$

This way one obtains a propagator

$$D^{R,2} = \left(F_0 - \frac{i m_F}{\sqrt{2}}\right)^{-1}. \tag{II.96}$$

Next one introduces a vertex $V^{(R,2)}(p, 1)$ describing the absorption (emission) of a scalar boson of momentum p out of a fermion line (fig. 8)

$$V^{(R,2)}(p, 1) = g_R \Gamma_5 :\mathrm{e}^{-ipQ(1)}: \tag{II.97}$$

Here, the new operator

$$\Gamma_5 = \gamma_5 (-1)^{\sum_{n=1}^{\infty} d_n^+ d_n} \tag{II.98}$$

is the generalization of the γ_5 matrix as it anticommutes with $\Gamma_\mu(z)$ and g_R is again a universal coupling constant. This choice for the vertex was quite inescapable to get gauge conditions in the $\mathfrak{F}_2$-picture made up from F_n-operators. The amplitude shown in fig. 8

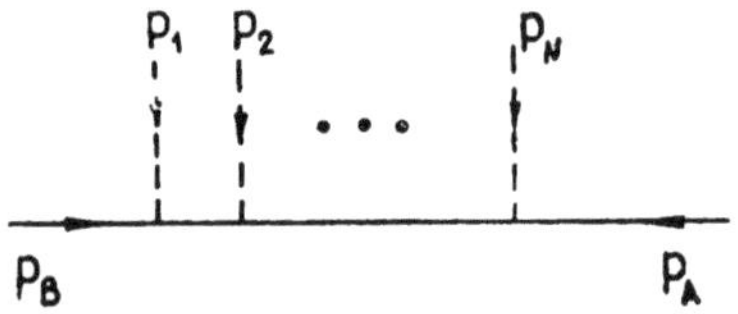

Fig. 8. Fermion-meson amplitude of the NSR-model

may now be written down [35] by sandwiching a sequence of propagator and vertex operators (II.96), (II.97) between the fermion ground states $|0, p_A\rangle\, u(p_A)$ and $\bar{u}(-p_B)\, \langle 0; -p_B|$ ($u(p_A)$ is the usual 2^{D-2}-dimensional free Dirac spinor) just as has been done in earlier cases. Furthermore, one may introduce in analogy to eq. (II.81) the Virasoro operators

$$L_n^R = L_n^a + L_n^d$$

$$L_n^d = \left(\frac{1}{i\sqrt{2}}\right)^2 \frac{1}{2} \left\langle z^n \left(:\Gamma(z)\, z \frac{d}{dz}\, \Gamma(z): - \frac{D}{4} \right) \right\rangle. \tag{II.99}$$

The term $-D/4$ is a zero mode correction for L_0^d

$$L_0^d = \frac{D}{16} + H^d = \frac{D}{16} - \sum_{n=1}^{\infty} n\, d^+_n\, d_n \tag{II.100}$$

due to the zero mode γ-matrices so that the projective algebra is maintained ($n = 0, \pm 1$). The Virasoro algebra of the L_n^R operators comes out to be identical to that of the L_n^{NS} operators eq. (II.82) so that one gets the same critical space-time dimension $D_c = 10$ as in the NS-model. Again close to the NS-case one finds the following commutators between F_n- and L_n^R-operators and anticommuntators between F_n-operators (comp. eq. (II.86), $-G_k \to F_n$)

$$\begin{aligned} [L_m^R, F_n] &= \left(\frac{m}{2} - n\right) F_{m+n}, \\ \{F_m, F_r\} &= 2L^R_{m+n} + \frac{D}{2}\left(m^2 - \frac{1}{4}\right) \delta_{m,-n}. \end{aligned} \tag{II.101}$$

The last equation leads after multiplying the Ramond-Dirac equation (II.95) by $F_0 + im_F/\sqrt{2}$ from the left to the corresponding 'Klein-Gordon' wave equation defining the mass spectrum

$$\begin{aligned} \left(F_0^2 + \frac{m_F^2}{2}\right) |\psi^{R,2}_{\mathrm{on}}\rangle &= \left(L_0^R - \frac{D}{16} + \frac{m_F^2}{2}\right) |\psi^{R,2}_{\mathrm{on}}\rangle \\ &= \left(-\frac{p^2}{2} + \frac{m_F^2}{2} + H^a + H^d\right) |\psi^{R,2}_{\mathrm{on}}\rangle = 0. \end{aligned} \tag{II.102}$$

As $H^a + H^d$ has integer eigenvalues $l^a + l^d$ the leading fermion trajectory with maximal spin $j = 1/2 - l^a + l^d$ is now given by

$$\alpha_F(p^2) = \frac{1}{2} + \frac{p^2 - m_F^2}{2}. \tag{II.103}$$

II.2.4. The $\mathfrak{F}_1$-reformulation and gauges for the Ramond fermion model

One may derive a modified operator expression for the amplitude of fig. 8 if one rewrites the propagator by the use of eq. (II.101) as

$$\frac{1}{F_0 - \frac{im_F}{\sqrt{2}}} = \frac{F_0 + \frac{im_F}{\sqrt{2}}}{L_0^R - \frac{D}{16} + \frac{m_F^2}{2}} \tag{II.104}$$

and then pushes all $(F_0 + im_F/\sqrt{2})$-operators in the nominator through the vertices until they can trivially be absorbed by a vacuum state. The $\mathfrak{F}_1$-representation of the $N + 2$-point function in fig. 8 emerges after all these manipulations in the following form

$$B^{R,1}_{N+2}(\bar{B}, 1, 2, \ldots, N, A) = \bar{u}(-p_{\bar{B}}) \langle 0; -p_{\bar{B}}| V^{(R,1)}(p_1, 1) \frac{1}{L_0^R - \frac{D}{16} + \frac{m_F^2}{2}}$$
$$\times V^{(R,1)}(p_2, 1) \frac{1}{L_0^R - \frac{D}{16} + \frac{m_F^2}{2}} \cdots \frac{1}{L_0^R - \frac{D}{16} + \frac{m_F^2}{2}} V^{(R,1)}(p_N, 1) |0; p_A\rangle u(p_A). \tag{II.105}$$

Here, the new vertex $V^{(R,1)}$ is given by

$$V^{(R,1)}(p, 1) = \{F_0, V^{(R,2)}(p, 1)\} = g_R p \cdot \Gamma(1) \sum \Gamma_5 :e^{-ipQ(1)}: \tag{II.106}$$

while the $\mathfrak{F}_1$-propagator as appearing in eq. (II.105) is now defined by a projective generator (L_0^R) so that the duality properties of the amplitude become more transparent in the $\mathfrak{F}_1$-picture. It turns out that $V^{(R,1)}$ has again a definite conformal spin $J_{V(R)} = -1/2 + p^2/2$ due to the conformal spin $-1/2$ of $\Gamma_\mu(z)$. To have the L_n^R operators as gauges, $J_{V(R)}$ has to be (-1) again, and this way the Ramond vertex describes necessarily the absorption (emission) of a spinless particle with mass squared $m^2 = -1$, i.e. this particle hast just the mass of the NS-pion (comp. eq. (II.83)). Besides the L_n^R gauges there are now additional F_n gauges to eliminate the ghosts related to the d^+_{n0} modes. In particular, the spectrum of the physical fermion states turns again out to be ghost-free provided that $D \leqq 10$ and the ground state fermion mass m_F is zero [*36*].

II.2.5. Fermion emission vertex

It was found out by Neveu, Schwarz and Thorn [*35*] in a relatively early stage of the development of the dual fermion model already that it is just the NS-meson spectrum that couples to af fermion-antifermion ($f\bar{f}$) ground state pair. Thus, the NS-meson and Ramond fermion models proved out to be just two different sectors of a single model, which we refer to as the Neveu-Schwarz-Ramond (NSR) model in the following. Based

on the work of these authors, CORRIGAN and OLIVE [37] derived the vertex operator V_A describing the absorption (emission) of an excited fermion, called A (fig. 9a). Their starting point was the projectively variant $\mathfrak{F}_1$-representation (II.105).
There projective transformations corresponding to the successive 'dualizations' indicated in fig. -9b—c were performed in the integrand so that the operatorial form of V_A could

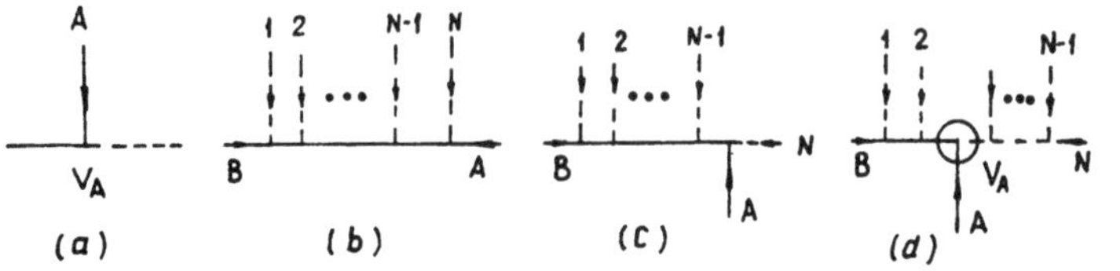

Fig. 9. (a) the Fermion emission vertex V_A, (b)—(d) successive factorization steps leading to the definition of V_A

be read off. After some lengthy calculations they obtained the following explicit form of $V_A(p_A, z)$

$$V_A(p_A, z) = g_R e^{-zL_{-1}^d} :e^{-ip_A Q(z)}: z^{1/2 p_A^2} \langle 0_b | \, e^{I_{+1}(z)+I_0(z)} e^{I_{-1}(z)} e^{E(z)} | A_d \rangle . \tag{II.107}$$

The operators $I_i(z)$ in the exponentials are given by

$$I_i = - \sum_{n \in N(i)} \sum_{k=1/2}^{\infty} d_n \gamma_5 b_k z^{-n-k} \left[(-1)^{n-k} \binom{-n - 1/2}{k - 1/2} \right] \tag{II.108}$$

with the summation ranges $N(i)$

$$N(i): \begin{cases} n \leqq -1 & \text{for} \quad i = +1 \\ n = 0 & \text{for} \quad i = 0 \\ n \geqq 1 & \text{for} \quad i = -1 . \end{cases}$$

The exponent E is a quadratic operator in terms of b-modes

$$E(z) = \frac{1}{2} \sum_{k,l=1/2}^{\infty} :b_k \, b_l: A_{kl}(z) \tag{II.109}$$

and the matrix A_{kl} is defined by the generating function

$$\sum_{k,l=1/2}^{\infty} x^{k-1/2} A_{kl}(z) \, y^{l-1/2} = \frac{1}{y - x} \left\{ 1 - \frac{1}{2} \sqrt{\frac{y - z}{x - z}} - \frac{1}{2} \sqrt{\frac{x - z}{y - z}} \right\} . \tag{II.110}$$

The vertex $V_A(p, z)$ depending on all three sets of a, b, d operators was derived in the $\mathfrak{F}_1$-picture. There exists a relation which is useful when working in the $\mathfrak{F}_2$-picture [38]

$$\langle R, 2 | \, V_A(p_A, 1) \, \gamma_5 u(p_A) \, | \text{NS}, 2 \rangle = - \frac{1}{\sqrt{2} \, m_F} \langle R, 1 | \, V_A(p_A, 1) \, u(p_A) \, | \text{NS}, 1 \rangle \tag{II.111}$$

where $|R, 1(2)\rangle$, $|\text{NS } 1(2)\rangle$ denote Ramond fermion and NS-meson states in the $\mathfrak{F}_1$-($\mathfrak{F}_2$-) picture.
The vertex (II.107) has been used to calculate the amplitude for fermion-antifermion scattering [39]. In a $(f\bar{f})$-channel one has to ensure now that for $D = 10$ the transverse (DDF-)meson states saturate any residuum. This has been proved to be the case if one

uses for the meson states between two ($f\bar{f}$)-states a 'projected' propagator [40]

$$\tilde{D}^{NS,2} = \int_0^1 \frac{dx}{x \cdot \Delta(x)}\, x^{L_0^{NS}-1/2}; \qquad \Delta(x) = (1-x)^{1/4}. \tag{II.112}$$

Concluding this section, let us mention that there are yet some weak points of the present dual fermion model, like parity doubling for all states except the ground state, absence of any three quark structure for the fermion channels due to the failure of the Chan-Paton procedure, absence of exotic channels present in baryon-baryon scattering. Thus, there arise difficulties if one wants to identify the Ramond fermions with baryons. As the spin couplings provided by the model agrees to a large extent with those of Feynman diagrams and the kinematical factors ensure automatically leading Regge behaviour it seemed nevertheless worthwile to test its phenomenological applicability e.g. in calculating density matrix elements of polarized particles (see sect. II.3.1.).

II.2.6. Some examples for scattering amplitudes in the NSR-model

Disregarding the insufficiencies of the NSR-model in describing realistic meson and baryon amplitudes we give here some examples for fermion-meson scattering where the meson and fermion ground states are called 'π' and 'N'.

i) $\pi N \to \pi N$ amplitude.
Let us consider the (s, t) and (s, u) diagrams of fig. 10a–b.

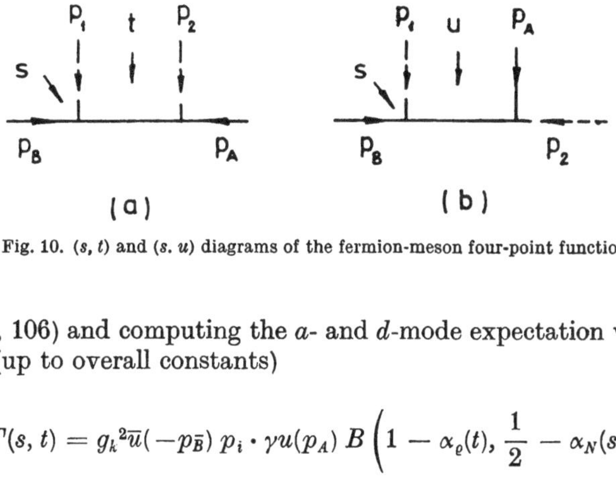

Fig. 10. (s, t) and $(s. u)$ diagrams of the fermion-meson four-point function

Using eq. (II.105, 106) and computing the a- and d-mode expectation values separately we easily obtain (up to overall constants)

$$T(s, t) = g_k^2 \bar{u}(-p_{\bar{B}})\, p_i \cdot \gamma u(p_A)\, B\left(1 - \alpha_\varrho(t), \frac{1}{2} - \alpha_N(s)\right) \tag{II.113}$$

where $\alpha_\varrho(t) = 1 + \alpha' t$, $\alpha_N(s) = 1/2 + \alpha'(s - m_\mp^2)$.
The amplitude of fig. 10b may conveniently be calculated in the $\mathfrak{F}_2$-picture. Taking the meson and fermion emission vertices (II.97), (II.107) one obtains with eq. (II.111)

$$T(s, u) = g_R^2 \bar{u}(-p_B) \langle 0; -p_B|\, V^{(R,2)}(p_1, 1)\, \frac{F_0 + \dfrac{i m_F}{\sqrt{2}}}{L_0^R - \dfrac{D}{16} + \dfrac{m_F^2}{2}}\, V_A(p_A, 1)\, \gamma_5 |0, p_2\rangle\, u(p_A). \tag{II.114}$$

Now we move the nominator of the propagator to the left and use eq. (II.106). This leaves us with the following a- and d-mode expectation values

$$\langle O_a| :e^{-ip_1Q(1)}: x^{L_0^a+(m_F^2/2)-1} :e^{-ip_AQ(1)}: |O_a\rangle = x^{[1/2-\alpha_N(s)]-1} (1-x)^{[1/2-\alpha_N(u)]+\alpha' m_\pi^2} \quad (II.115)$$

and

$$\langle O_d|\, p_1\Gamma\left(\frac{1}{x}\right) \Gamma_5 e^{L_{-1}^d}|O_d\rangle = (p_i\gamma)\,\gamma_5(1-x)^{-1/2} \quad (II.116)$$

where we have used the relation [37]

$$\Gamma_\mu(y)\, e^{zL_{-1}^d} = e^{zL_{-1}^d} \sqrt{\frac{y}{y-z}}\, \Gamma_\mu(y-z). \quad (II.117)$$

This yields finally

$$T(s,u) = g_R^2\bar{u}(-p_{\bar{B}})\,(p_1\cdot\gamma)\, u(p_A)\, B\left(\frac{1}{2}-\alpha_N(x),\, \frac{1}{2}-\alpha_N(u)\right). \quad (II.118)$$

ii) $\pi N \to \pi\pi N$ amplitude.

Let us now calculate as a further example the more complicated diagram of fig. 11.

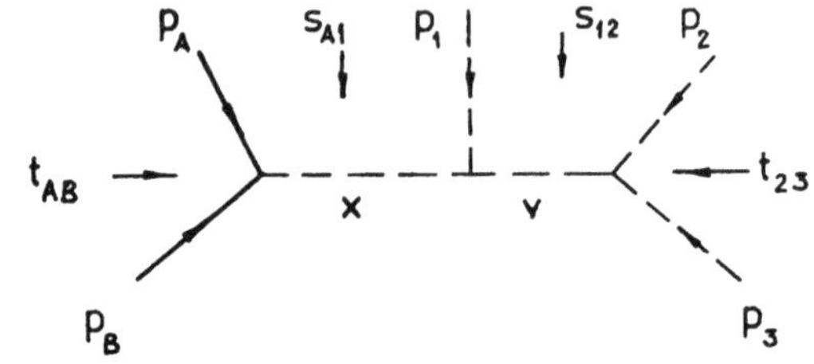

Fig. 11. Five-point function of the process $\pi N \to \pi\pi N$

Using the operatorial rules of the preceeding section we obtain in the $\mathfrak{F}_2$-picture

$$T = \bar{u}(-p_{\bar{B}})\,\langle 0_{a,d}|\, V_A(p_A,1)\,\gamma_5 u(p_A)\, \frac{1}{L_0^{NS}-\frac{1}{2}}\, V^{(NS)}(p_1,1)$$

$$\times \frac{1}{L_0^{NS}-\frac{1}{2}}\, V^{(NS)}(p_2,1)\,|0_{a,b};p_3\rangle. \quad (II.119)$$

The vacuum expectation value of the a-operators may be calculated in the standard way, whereas the b-mode expectation value is given by

$$\langle 0_b|\, V_A^{(b)} x^{L_0^b} H^\mu(1)\, y^{L_0^b} H^\nu(1)\,|0_b\rangle$$

$$= -\frac{1}{2}\, g^{\mu\nu}\, \frac{\sqrt{y}}{1-y}\left[\sqrt{\frac{1-x}{1-xy}}+\sqrt{\frac{1-xy}{1-x}}\right] + \frac{1}{4}\sqrt{\frac{x^2y}{(1-xy)\,(1-x)}}\,[\gamma^\mu,\gamma^\nu]. \quad (II.120)$$

In the last equation we have used eq. (II.110). We thus arrive at the final result

$$T = \frac{1}{2}\, g_{NS}^2 g_R \bar{u}(-p_B)\,\gamma_5 u(p_A)\cdot \alpha_\varrho(s_{12})$$

$$\times \left\{B_5\left(-\alpha_\pi(t_{AB}),\, 1-\alpha_\varrho(t_{23}),\, \frac{3}{2}-\alpha_N(s_{A1}),\, -\alpha_\varrho(s_{12}),\, \frac{1}{2}-\alpha_N(s_{A12})\right)\right.$$

$$+ B_5\left(-\alpha_\pi(t_{AB}), 1 - \alpha_\rho(t_{23}), \frac{1}{2} - \alpha_N(s_{A1}), -\alpha_\rho(s_{12}), \frac{1}{2} - \alpha_N(s_{A12})\right\}$$

$$+ \frac{1}{4}\, g_{NS}^2 g_R \bar{u}(-p_B)\, [\gamma^\mu, \gamma^\nu]\, (k_1)_\mu\, (k_1)_\nu\, \gamma_5 u(p_A)$$

$$\times B_5\left(1 - \alpha_\pi(t_{AB}), 1 - \alpha_\rho(t_{23}), \frac{1}{2} - \alpha_N(s_{A1}), 1 - \alpha_\rho(s_{12}), \frac{1}{2} - \alpha_N(s_{A12})\right) \quad \text{(II.121)}$$

where B_5 denotes the usual Veneziano five-point function written as functional of the trajectories of the different Mandelstam channels in fig. 11 and $\alpha_\pi(t) = 1/2 + \alpha' t$.

iii) $\pi N \to \rho N$ and $\pi N \to \omega N$ amplitudes.

These amplitudes may be calculated by factorizing B_5- and B_6-functions of the NSR-model in two- or three-meson channels. As the calculation is straightforward we quote only the results.

$$T_{\pi N \to \rho N}(s, t) = g_R \cdot g_{NS} \bar{u}(-p_B) \left\{-\frac{1}{2}\, p_1 \cdot \varepsilon(p_2)\right.$$

$$\times \left[B\left(-\alpha_\pi(t), \frac{3}{2} - \alpha_N(s)\right) + B\left(-\alpha_\pi(t), \frac{1}{2} - \alpha_N(s)\right)\right]$$

$$\left. + \frac{1}{4}\, p_1{}^\nu \varepsilon^\mu(p_2)\, [\gamma_\nu, \gamma_\mu]\, B\left(1 - \alpha_\pi(t), \frac{1}{2} - \alpha_N(s)\right)\right\} \gamma_5 u(p_A) \quad \text{(II.122)}$$

and

$$T_{\pi N \to \omega N}(s, t) = g_R \cdot g_{NS} \bar{u}(-p_B) \cdot \frac{i}{\sqrt{2}}\, (p_1)_\rho\, (p_2)^\nu\, \varepsilon_\mu(p_2) \cdot \varepsilon^\mu_{\nu\lambda\sigma}$$

$$\times \left\{-\frac{1}{2}\, K_1{}^{\lambda\sigma\varrho} B\left(2 - \alpha_\rho(t), \frac{1}{2} - \alpha_N(s)\right)\right.$$

$$\left. + K_2{}^{\lambda\sigma\varrho}\left[-B\left(1 - \alpha_\rho(t), \frac{1}{2} - \alpha_N(s)\right) + \frac{1}{2}\, B\left(2 - \alpha_\rho(t), \frac{1}{2} - \alpha_N(s)\right)\right]\right\} u(p_A) \quad \text{(II.123)}$$

where

$$K_1{}^{\lambda\sigma\rho} = -\frac{i}{6}\, \gamma^5 \gamma^\alpha \varepsilon_\alpha{}^{\lambda\sigma\rho} \qquad K_2{}^{\lambda\sigma\rho} = \gamma^\sigma \gamma^{\rho\lambda} - \gamma^\lambda g^{\rho\sigma} \quad \text{(II.124)}$$

and $\varepsilon_\mu(p)$, $\varepsilon^\mu_{\nu\lambda\sigma}$ are the spin-one polarization vector and the Levi-Civita tensor, respectively. The coupling of the π-trajectory to the $(\pi\rho)$ and $(N\bar{N})$ states in the t-channel is determined by the residue of the amplitude (II.122) at $\alpha_\pi(t) = 0$. We obtain

$$\beta_\pi = -g_{NS} \cdot g_R\big(p_i \cdot \varepsilon(p_2)\big)\, \bar{u}(-p_B)\, \gamma_5 \cdot u(p). \quad \text{(II.125)}$$

Eq. (II.125) agrees with the usual coupling scheme of a pseudoscalar particle. On the other hand, the coupling of the ω-trajectories is given by the residuum of the same amplitude at $\alpha_\pi(t) = 1$. One obtains

$$\beta_\omega = -g_{NS} \cdot g_R\big(p_1 \varepsilon(p_2)\big)\, \alpha_N(s)\, \bar{u}(-p_B)\, \gamma_5 u(p_A)$$

$$+ \frac{1}{2}\, g_{NS} g_R p_1{}^\nu \varepsilon^\mu(p_2)\, \bar{u}(-p_B)\, [\gamma_\nu, \gamma_\mu]\, \gamma_5 u(p_A). \quad \text{(II.126)}$$

Using the conditions $\alpha' m_\pi^2 = -1, 2\ m_\rho = 0, \alpha' m_\omega^3 = 1/2$ this expression may also be written in the form of a 'magnetic' coupling of a $j^P = 1^-$-particle to the $f\bar{f}$-system

$$\beta_\omega = \frac{1}{2} g_{NS} g_R \varepsilon_{\varrho\varkappa\lambda\sigma}(p_2)^\varkappa \varepsilon^\lambda(p_2)\,(p_1)^\sigma\,\bar{u}(-p_{\bar{B}})\,(p_A - p_{\bar{B}})^\nu\,[\gamma_\nu, \gamma^\varrho]\,u(p_A). \quad \text{(II.127)}$$

Contrary to this, the ρ-meson of the NSR-model couples only 'electrically' via $\bar{u}(-p_{\bar{B}})$ $\gamma_\mu u(p_A)$.

II.3. Phenomenological Applications

In this section we want to give a short survey on some phenomenological applications of the DRM to the description of exclusive ('all final state particles are measured') and inclusive experiments.

It is typical for almost all of these applications that one introduces physical trajectories with an imaginary part $\text{Im}\,\alpha = \alpha' \cdot m_{\text{res}} \cdot \Gamma_{\text{res}}$ into the model tolerating thus the appearance of ancestors and ghosts which are generally believed to be unimportant in the respective kinematical regions. There is a further problem connected with the Pomeron singularity. This singularity appears as is well known only in vacuum channels and cannot be simply described in the framework of a dual tree amplitude as it is dual to a background and not to a resonance behaviour. To overcome this difficulty one has either to exclude it by considering reactions without vacuum channels or should include suitable defined dual Pomeron-background amplitudes. Despite of such difficulties the phenomenological applications of the DRM have the attractive feature that the resultant amplitudes contain only very few parameters and offer a unified description of phenomena previously regarded as separate ones as, for example, resonance production and multi-Regge behaviour. A further advantage is that the resultant model amplitudes may describe a great number of different reactions related by crossing.

II.3.1. B_5-phenomenology for exclusive reactions

The Veneziano B_5-function has been first successfully applied to reactions of the type $(\bar{K}\pi\pi N/)$ and $(\bar{K}K\pi N\bar{N})$ [*41, 42*]. For example, the reactions

i) $K^+p \to K^0\pi^+p$

ii) $K^-p \to \bar{K}^0\pi^-p$

iii) $\pi^-p \to K^0\bar{K}^-p$

are supposed to be described, due to charge and crossing symmetry, by the same dual amplitude. This amplitude is given as a linear combination of twelve B_5-functions corresponding to the twelve nonequivalent permutations of the external particles. Among these there remain only four B_5-functions describing configurations without exotic channels. A typical contributing term corresponding to the graph of fig. 12 is [*42*]

$$T = \varepsilon_{\mu\nu\varrho\sigma} p_\pi{}^\mu p_{K'}{}^\nu p_K{}^\varrho p_p{}^\sigma B_5\left(1 - \alpha_{K^*}, 1 - \alpha_A, \frac{1}{2} - \alpha_\Lambda, 1 - c\alpha_\omega, \frac{1}{2} - \alpha_N\right). \quad \text{(II.128)}$$

In this ansatz the baryons are treated as spin zero mesons, furthermore, when several trajectories may couple, only the leading one is taken into account. The kinematical factor in eq. (II.128) ensures that the trajectory in every channel starts with a p-wave resonance. As has been found, this model describes satisfactorily well with only one free overall normalization parameter the energy dependence of the total cross section, the momentum transfer distribution and effective mass and decay angular distributions of the above reactions.

Despite of this rather good over-all agreement with data a more detailed description of the experiments, e.g. of polarization phenomena requires to take into account the spin structure of the reaction considered. Several attempts have been made into this direction with the aim to construct amplitudes for meson-baryon processes [*43*]. In these approaches the necessary kinematical factors multiplying the Veneziano function are

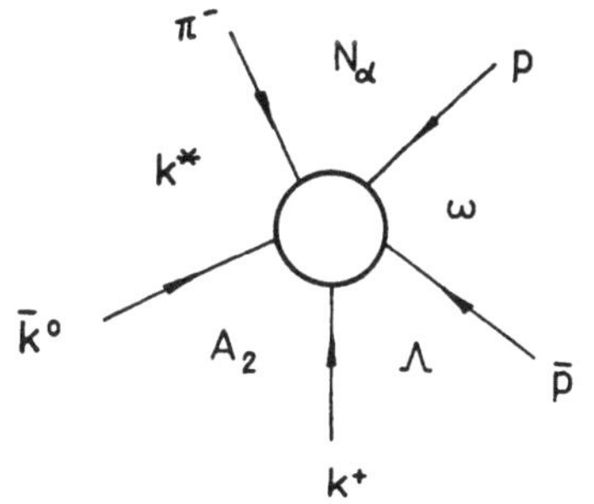

Fig. 12. Typical graph of ref. [*42*] contributing to the process (KKπNN)

adjusted by hand in such a way that one obtains correct Regge behaviour as well as a reasonable coupling structure for the particle poles in the different Mandelstam channels. As a further possibility one can use directly the more sophisticated NSR-model discussed in sect. II.2., at least as a guide, in order to construct scattering amplitudes with fermion spin. Then, the kinematical factors required to guarantee Regge behaviour and some spin-parity properties of the particles involved are fixed within the model.

Let us consider as an example the quasi-three-body reaction $K^-p - \overline{K}^{*0}$ (890) (π^-p) at 10 GeV/c investigated in ref. [*44*]. Due to the exoticity criteria there remain the three relevant dual diagrams of fig. 13a—c that contribute to the helicity amplitudes with definite parity exchange in the t-channel (since the separation into natural and unnatural spin-parity exchange holds to leading order in the energy squared, the single Regge limit of the amplitude has been taken).

Treating the kaons only with regard to their spin-parity content, the graphs of fig. 13a—c have been calculated by factorizing six-point meson-fermion amplitudes of the NSR-model on the spin 1 resonance $(\overline{K}^{*0})$ of the two-meson (23)-channel (comp. fig. 13c). The helicity amplitude is then given in the single Regge limit by

$$H_\lambda = \varepsilon_\mu(\lambda)\, \Gamma\left(1 - \alpha^\pi_{123}\right) \left[\left(-\alpha^N_{16}\right)^{\alpha^\pi_{123}} B^\mu_{\bar{s}\bar{t}} + \left(-\alpha^N_{146}\right)^{\alpha^\pi_{123}} B^\mu_{\bar{t}\bar{u}} + \left(-\alpha^N_{16}\right)^{\alpha^\pi_{123}} B^\mu_{\bar{s}\bar{u}}\right] \tag{II.129}$$

where $\varepsilon_\mu(\lambda)$ is the polarization vector of the $\overline{K}^{*0}$ and the trajectories are labeled as $\alpha^r_{ij} = \alpha^r(s_{ij}) = \alpha_0{}^r + \alpha' s_{ij}$ with $s_{ij} = (k_i + k_{i+1} + \cdots + k_j)^2$, r denoting the type of the Regge trajectory. The functions $B^\mu_{\bar{s}\bar{t}}$ etc. $\left(\bar{s} = (p_4 + p_5)^2, \bar{t} = (p_5 + p_6)^2, \bar{u} = (p_4 + p_6)^2\right)$ are given by a sum of different terms characterizing the K^-K^{*0}-Reggeon-vertex and the four point amplitude with an external Reggeon (comp. fig. 13). It turns out that all these expressions have by construction the right shifts of fermion and meson trajectories by half-integers or integers, respectively, and, due to their kinematical factors, also the correct Regge behaviour. One has finally to give some additional prescription in order to overcome the unphysical π—ω-degeneracy of the NSR-model and to be able to insert a physical π-trajectory. For this, one considers helicity amplitudes with definite parity exchange in the t-channel, $H_\lambda{}^\pm = H_\lambda \mp (-1)^\lambda H_{-\lambda}$ and relates the unnatural spin-parity trajectory (π) to H_{λ}- and the trajectory with natural spin-parity (ω) to H_{λ}^{+}. Furthermore, one needs yet some modification to include besides the N_α-trajectories also the Δ- and N_γ-recurrences [44]. Fig. 14a–b shows the comparision of the model predictions with

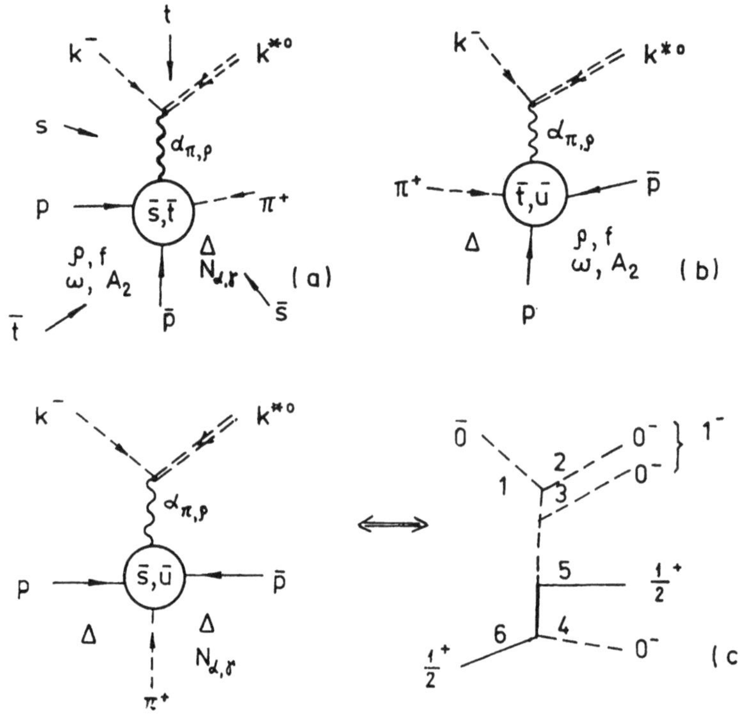

Fig. 13. Single Regge limit of the different dual configurations contributing to the process $K^-p \to K^{*0}(\pi p)$. The last graph represents a corresponding dual tree diagram of the NSR-model

the data. The dashed curves for the unnatural exchange cross sections correspond to single pion-exchange whereas the solid curves have been obtained after including also a B-trajectory. One then obtains reasonable results for the differential cross sections, the effective mass distributions and the K^{*0} density matrix elements. Similar investigations have been also performed for the quasi-two-body processes $\pi^-p \to (\rho^0, \omega)\,n$, $K^-p \to \overline{K}^*$ (890, 1420) N [45].

II.3.2. B_6, B_8-phenomenology for inclusive reactions

Let us consider the inclusive reaction

$$a + b \to c + X \text{ (anything)}. \tag{II.130}$$

A suitable set of variables to describe this reaction in the case of spinless particles is the square of the cms-energy $s = (p_a + p_b)^2$ and the longitudinal and transverse momentum $p_\parallel$, $\mathfrak{p}_\perp$ of the particle c (defined with respect to the beam direction). Other useful sets are

$$\begin{aligned} &\text{i) } s,\ \mathfrak{p}_\perp \text{ and the Feynman variable } x = 2p_\parallel^{\text{cms}}/\sqrt{s} \\ &\text{ii) } s,\ \mathfrak{p}_\perp \text{ and the rapidity } y \text{ defined by } p_\parallel = \sqrt{p_\perp^2 + m_c^2}\cdot\sinh(y) \\ &\text{iii) Mandelstam variables } s,\ t,\ u,\ M^2 \text{ (defined as in fig. 15).} \end{aligned} \tag{II.131}$$

In the high energy limit ($s \to \infty$) the following four kinematical regions are of particular interest (the particle b is supposed to move in the positive z direction)

a) fragmentation region of $b (x > 0)$

$$-u = xs \qquad M^2 = (1 - x)\, s$$

$$-t = -m_c^2 - m_b^2(1 - x) + \frac{\mu^2}{x}, \qquad \mu^2 = m_c^2 + p_\perp^2;$$

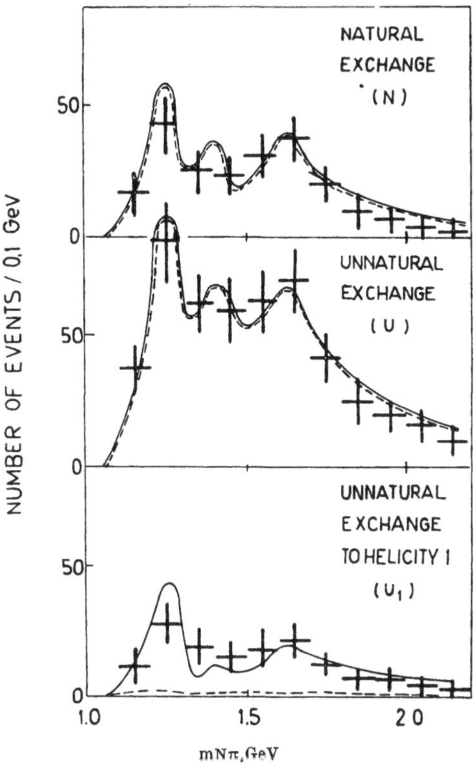

Fig. 14a. The natural, the total unnatural and the unnatural exchange contributions to helicity 1 as functions of the nucleon-pion mass (taken from ref. [44])

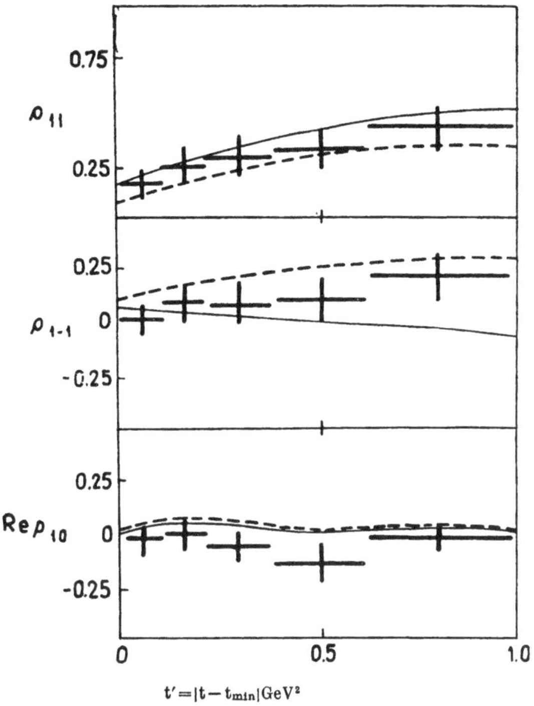

Fig. 14b. The K*⁰ (890) density matrix elements ρ_{11}, ρ_{1-1} and Re ρ_{10} as functions of $t' = |t - t_{\min}|$

b) triple Regge region ($|x| \to 1$)
This is a special case of the fragmentation region with

$$s \gg M^2 \gg 1\,\mathrm{GeV}^2, \text{ i.e. } \frac{M^2}{s} \to 0;$$

c) pionization region ($x \approx 0$)

$$|u| \approx |t| = O(s^{1/2}) \quad \text{with} \quad \frac{tu}{s} = \mu^2;$$

d) fixed angle limit

$$\left.\begin{array}{lll} t \sim -\frac{s}{2}\, r(1-z) & M^2 = s(1-r) & \\ u \sim -\frac{s}{2}\, r(1+z) & r = \frac{|\boldsymbol{p}_c|}{|\boldsymbol{p}_a|}, & z = \cos\theta \end{array}\right\} \qquad \text{(II.132)}$$

(θ designates the angle between the particles a and c).

The differential cross section of reaction (II.130) may be related through the generalized optical theorem [46] to the discontinuity of the three-particle elastic forward scattering amplitude calculated in the missing mass channel according to

$$(2\pi)^3\, 2p^0\, \frac{d^3\sigma_{ab,c}}{d^3\boldsymbol{p}}\, 2\sqrt{\lambda(s, m_a^2, m_b^2)} = \mathrm{disc}_{M^2}\, T_{33}(M^2, s + i\varepsilon, s' - i\varepsilon, \ldots) \qquad \text{(II.133)}$$

where $2\lambda^{1/2}$ is the flux factor of the incident particles. Using the B_6-function as a model for T_{33} one may investigate by eq. (II.133) the asymptotic behaviour of the differential cross section [47–50]. Thus, one has found scaling behaviour (as $s \to \infty$ the s-dependence in the cross section drops out leaving us only with a dependence on $p_\perp$ and the scaled variable x) for the differential cross section normalized by $\sigma_{ab} \sim s_{\alpha_{01}}/2\lambda^{1/2}$. For example, one obtains in the triple Regge region (comp. eq. II.132b) [50]

$$\mathrm{disc}_{M^2}\, B_6 \xrightarrow[x\to 1]{s\to\infty} \frac{(\alpha' s)^{\alpha_{01}}}{\Gamma(1+\alpha_{01})}\, (1-x)^{\alpha_{01} - \alpha_{23}(t) - \alpha_{45}(t)}\, \frac{\Gamma(-\alpha_{23}(t))\, \Gamma(-\alpha_{45}(t))\, \Gamma(1+\alpha_{01}(0))}{\Gamma(1 - \alpha_{01}(0) - \alpha_{23}(t) - \alpha_{45}(t))} \qquad \text{(II.134)}$$

which corresponds to the three-Reggeon graph of fig. 15.

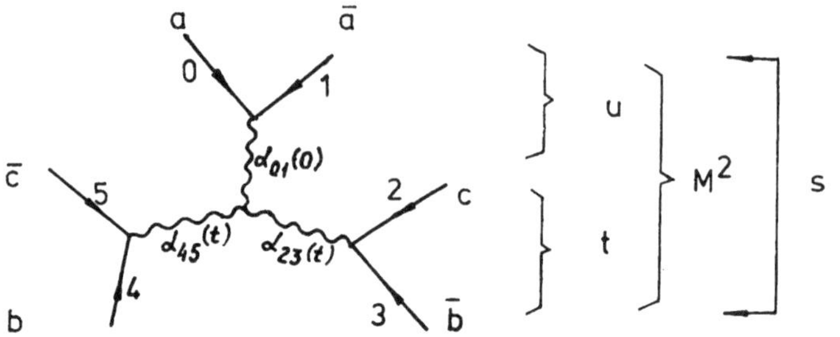

Fig. 15. Three-Reggeon graph

Furthermore, the following high $p_\perp$-behaviour may be derived in the fragmentation region (eq. (II.132a) [50]

$$\mathrm{disc}\, B_6 \xrightarrow[p_\perp^2\to\infty]{} \left[(\alpha' s)^{\alpha_{01}}\, x^{\alpha_{01}-\alpha_{34}}\, \sqrt{\pi}\, (\alpha'\mu^2)^{-\alpha_{01}-3/2}\right] \left[\left(\frac{1-x}{1+x}\right)^{\alpha_{01}-2\alpha(t)}\right]$$

$$\times \left[\left(\frac{1+x}{2}\right)^{2(\alpha_{01}+\alpha_{34}+1) - \alpha_{03}(m_c^2) - \alpha_{51}(m_c^2)}\right]. \qquad \text{(II.135)}$$

Note that the second factor in eq. (II.135) leads to an exponential decrease in $p_\perp^2$ with the slope

$$\lambda(x) = 4\alpha' \cdot \frac{1}{2x} \ln \frac{1+x}{1-x} \qquad \text{(II.136)}$$

which takes the value $\tau(0) = 4\alpha' \simeq 4\ (\text{GeV}/c)^{-2}$ in the pionization region ($x \approx 0$). Thus, the DRM supplies a universal transverse momentum cut-off $\exp\left[-4\alpha'\mu^2\right]$ which yields for the average values the bounds $\langle p_\perp^2 \rangle \leqq (4\alpha')^{-1} \simeq 0.25\ (\text{GeV}/c)^2$ and $\langle |p_\perp| \rangle \leqq \sqrt{\pi}/2 \times (4\alpha')^{-1/2} \simeq 450$ MeV/c, in qualitative agreement with experiment.
HUANG and SEGRÈ [*51*] proposed within the DRM also the following inelastic scaling law of the single particle distribution in the fixed angle limit (II.132d).

$$\ln \left(\frac{d^2\sigma}{dt\, dM^2}\right)_{\text{c.m.}} \Big/ s \cdot G(r, z) \xrightarrow[s\to\infty]{} \alpha' \tag{II.137}$$

where

$$G(r, z) = (1 + rz) \ln (1 + rz) + (1 - rz) \ln (1 - rz) - (1 + r) \ln (1 + r) - (1 - r) \ln (1 - r). \tag{II.138}$$

The scaling law (II.137) is pretty well satisfied in the reactions $p + p \to (p, \pi^\pm) + X$ for momenta between 10–30 GeV/c ($|z| < 0{,}7$) although the value on the right-hand side is rather $\alpha'/2$. This deviation from the value α' of the tree amplitude may be a hint that one has to include additional dual Pomeron contributions [*52*]. A detailed numerical analysis for single particle distributions calculated within DRM can be found in ref [*53*]. Similar investigations have been performed for inclusive two-particle reactions

$$a + b \to c_1 + c_2 + X \tag{II.139}$$

using the dual B_8-function as a model amplitude [*54*]. As $s \to \infty$ one finds again scaling behaviour of the differential cross section. In particular, if $1 - x_1 - x_2 \equiv M^2/s \to 0$ $\left(x_i = 2p_{i\parallel}^{\text{c.m.}}/\sqrt{s},\ i = c_1, c_2\right)$ the triple Regge behaviour follows [*50*]

$$\operatorname{disc} B_8 = (\alpha' s)^{\alpha_{01}} (1 - x_1 - x_2)^{\alpha_{01} - 2\alpha(t_{b12})} \frac{\Gamma^2\left(-\alpha(t_{b12})\right)}{\Gamma\left(1 - \alpha_{01} - 2\alpha(t_{b12})\right)} \times \left[\left|B_4\left(-\alpha(s_{12}), -\alpha(t_{b2})\right)_2 F_1\left(-\alpha(t_{b12}) - \alpha(s_{12}), -\alpha(s_{12}) - \alpha(t_{b2}); \frac{x_2}{x_1 + x_2}\right)\right|^2\right] \tag{II.140}$$

where ${}_2F_1$ is the hypergeometric function. In the pionization region of the two particles c_1, c_2 ($x_1 \approx 0$, $x_2 \approx 0$) there is again a transverse momentum cut-off [*54*]

$$\operatorname{dis} B_8 \sim e^{-4\alpha'(p_{1\perp}^2 + p_{2\perp}^2 + 2p_{1\perp}p_{2\perp}\cos\beta)}. \tag{II.141}$$

If $p_{1\perp} \gg p_{2\perp} \gg m_1, m_2$ or $p_{1\perp} = p_{2\perp} \gg m_1, m_2$ the angle β appearing in eq. (II.141) is only a function of the rapidity difference $y_1 - y_2$ of the particles c_1, c_2 and of the azimuthal angle φ defined by $\cos\varphi = \boldsymbol{p}_{1\perp} \cdot \boldsymbol{p}_{2\perp}/|\boldsymbol{p}_{1\perp}| \cdot |\boldsymbol{p}_{2\perp}|$. It can be shown that $\cos\beta(y_1 - y_2, \varphi) \geqq \cos\beta(y_1 - y_2, \pi)$ so that the two particles are produced according to eq. (II.141) most frequently in opposite directions. Furthermore, the following expression of the two-particle correlation function has been derived

$$C^{(2)} \approx -\frac{32}{\alpha(0)} (p_{1\perp} p_{2\perp})^2 \left\{\frac{p_{1\perp}^2 + p_{2\perp}^2}{2p_{1\perp}p_{2\perp}} + \cos\varphi\right\} e^{-|y_1 - y_2|}. \tag{II.142}$$

Thus, the correlation length L defined by the behaviour $\exp\left(-|y_1 - y_2|/L\right)$ of the correlation function has in the tree approximation of the DRM the value one. Interesting results may also be derived for the production of vector mesons by factorizing a

B_8-function on the resonance poles in a two-particle subenergy channel. This yields then a B_6-function with an external vector meson which can be used to describe decay correlations and density matrix elements of inclusive reactions with vector mesons [55]. Furthermore, one has also investigated the influence of higher order loop amplitudes (comp. chapt. III) on inclusive reactions [56].

III. Investigation of Dual Loop Amplitudes

III.1. Introductory remarks

Up to now the tree graphs of the DRM have been discussed. As we have seen, they were meromorphic functions on the Mandelstam variables with pole singularities reflecting the propagation of intermediate one-particle states. Due to the unitarity equation the imaginary part of the physical scattering amplitude should include, however, contributions from many-particle intermediate states, too, leading to cuts in the Mandelstam variables. It seems thus to be important to consider, besides the tree graphs, also dual loop amplitudes containing more than one internal propagator line. Moreover, besides producing normal threshold singularities, these loop contributions should provide corrections to the 'bare' Regge trajectories of the tree amplitudes, i.e. to the spectrum. In close analogy to quantum electrodynamics, the tree graphs of the DRM may thus be viewed as the Born terms of a dual theory, and a unitary perturbation series should be constructed in which loop amplitudes appear as higher order corrections.

The loop graphs are usually classified into planar, nonplanar orientable or nonplanar nonorientable ones, respectively, depending on whether they contain no twisted internal lines or an even or odd number of twisted propagators [8]. In fig. 16a–b we exhibit typical examples of planar and nonplanar orientable one-loop four-point functions in different dual equivalent configurations.

III.2. General expressions for loop amplitudes

III.2.1. One-loop amplitudes in the conventional model

In this section we want to collect some one-loop formulae as well as the expressions of the one-loop self-energy operators. For this, let us recall, that the loops have been constructed by 'sewing' to excited legs of a tree amplitude with external Reggeons inserting a propagator between them [57] (comp. Fig. 16a)

Thus, the planar one-loop graph of fig. 16a turns out to be given by (we restrict us, for convenience, to the four-point case)

$$T_1 = \int d^Dk I(k; p_1, p_2, p_3, p_4) \tag{III.1}$$

with

$$I = \mathrm{Tr}\big[D(k^2)\, V(p_2)\, D\big((k - p_1)^2\big)\, V(p_2) \dots D\big((k + p_4)^2\big)\, V(p_4)\big]. \tag{III.2}$$

Performing the loop momentum integration as well as the trace calculation yields the following formula for the planar one-loop four-point amplitude [57, 58]

$$T_1^{\mathrm{pl}} = (2\pi)^{D/2} g^4 \int_0^1 \prod_{i=1}^{4} [dx_i(1 - x_i)^{a-1}]\,(1 - \omega)^F f^E(\omega)\, \omega^{-a-1} \frac{f(\omega)^{-D}}{(-\ln \omega)^{D/2}} \prod_{1 \leq i < j \leq 4} \psi(x_{ij})^{-p_ip_j} \tag{III.3}$$

where

$$\omega = \prod_{i=1}^{4} x_i \qquad x_{ij} = x_{i+1} \cdot x_{i+2} \ldots x_j \quad .$$

$$f(\omega) = \prod_{n=1}^{\infty} (1 - \omega^n). \tag{III.4}$$

Similarly, by inserting twisted propagators, the nonplanar amplitude of fig. 16b is given by[7])

$$T_1^{\text{npl}} = (2\pi)^{D/2} g^4 \int \prod_{i=1}^{4} [dx_i(1 - x_i x_{i+1})^{\alpha-1}] (1 - \omega)^G f^E(\omega) \, \omega^{-\alpha-1}$$
$$\times \frac{f(\omega)^{-D}}{(-\ln \omega)^{D/2}} \prod_{1 \leqq i < j \leqq 4} [\psi(x_{ij}) \quad \text{or} \quad \psi_T(x_{ij})]^{-p_i p_j}. \tag{III.5}$$

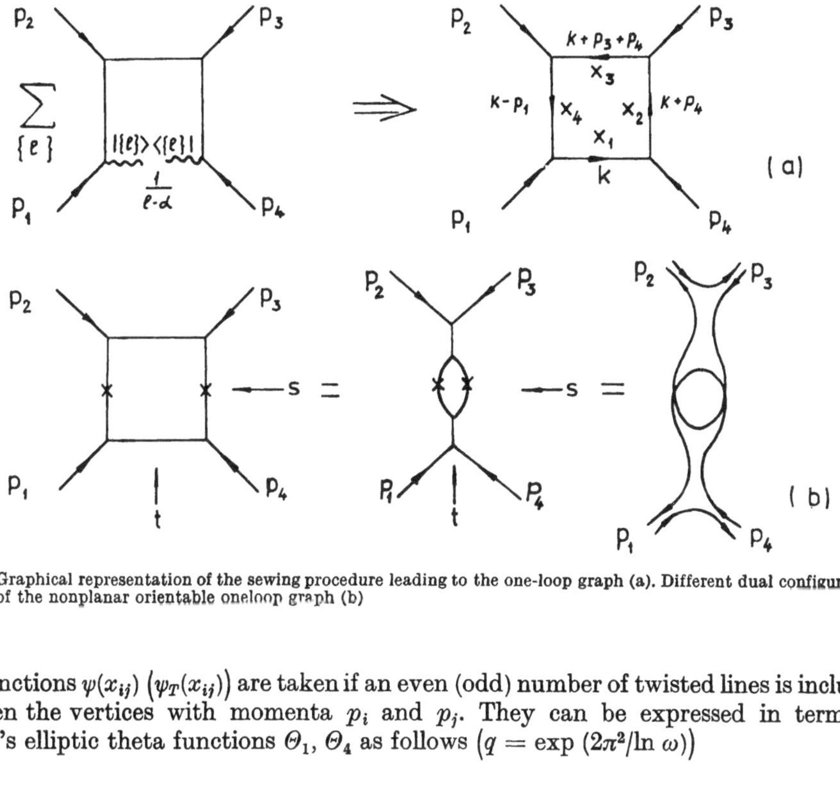

Fig. 16. Graphical representation of the sewing procedure leading to the one-loop graph (a). Different dual configurations of the nonplanar orientable oneloop graph (b)

The functions $\psi(x_{ij})$ $\left(\psi_T(x_{ij})\right)$ are taken if an even (odd) number of twisted lines is included between the vertices with momenta p_i and p_j. They can be expressed in terms of Jacobi's elliptic theta functions Θ_1, Θ_4 as follows $\left(q = \exp\left(2\pi^2/\ln \omega\right)\right)$

$$\psi(x, \omega) = -\ln \omega \cdot \Theta_1 \left(\frac{\ln x}{\ln \omega} \middle| -\frac{2\pi_i}{\ln \omega}\right) \Big/ \Theta_1' \left(0 \middle| -\frac{2\pi i}{\ln \omega}\right)$$

$$= -\frac{\ln \omega}{\pi} \sin\left(\frac{\pi \ln x}{\ln \omega}\right) \prod_{n=1}^{\infty} \frac{1 - 2q^{2n} \cos \frac{2\pi \ln x}{\ln \omega} + q^{4n}}{(1 - q^{2n})^2}, \tag{III.6}$$

[7]) The correct integration region will be given in terms of Jacobi transformed variables (see eq. III.11).

$$\psi_T(x, \omega) = -\ln \omega \Theta_4 \left(\frac{\ln x}{\ln \omega} \middle| -\frac{2\pi i}{\ln \omega}\right) \Big/ \Theta_1{}' \left(0 \middle| -\frac{2\pi i}{\ln \omega}\right)$$

$$= -\frac{1}{2} q^{-1/4} \ln \omega \prod_{n=1}^{\infty} \frac{1 - 2q^{2n-1} \cos \frac{2\pi \ln x}{\ln \omega} + q^{4n-2}}{(1 - q^{2n})^2}. \quad \text{(III.7)}$$

In eq. (III.3, 5) we have admitted additional factors $(1 - \omega)^{F,G}$, $f^E(\omega)$ which come essentially from the elimination of spurious states by inserting a projection operator on physical states inside the operator trace (III.2). If $a \neq 1$, $D = 4$ one finds, for example, $E = 0$, $F = 1$, $G = 2$ whereas in the case $a = 1$, $D = 4$ the values are $E = F = G = 1$ [59]. The appearance of the power E of the 'partition function' $f(\omega)$ reflects then just the decoupling of the complete set of timelike oscillator modes. BRINK, OLIVE and COLLINS, FRIEDMAN [60] recalculated the one-loop graphs in the critical dimension of space-time $(a = 1)$ using the physical state projection operator $\tau(k)$ introduced in ref. [61]. They obtained $E = 2$, $F = G = 0$. Then the whole set of longitudinal modes decouples, too, (comp. sect. II.1.7) and the previous factor $(1 - \omega)$ is converted into an additional factor $f(\omega)$. Taking into account only the transverse DDF-states leads then in the trace calculation to an effective power $-(D - 2)$ of the partition function.[8]) From reasons that will become clear below from the investigation of the nonplanar diagram we shall restrict ourselves in the following to the latter case $a = 1$, $D = D_{\text{crit.}}$.

The loop expressions (III.3, 5) may be rewritten in another suitable form by performing the Jacobi transformation wellknown in the theory of elliptic functions

$$\Theta = \frac{\pi \ln x}{\ln \omega}; \qquad q = \exp\left(\frac{2\pi^2}{\ln \omega}\right) \quad \text{(III.8)}$$

and using

$$\left(\frac{2\pi}{\ln \omega}\right)^{D/2} f^{-D}(\omega) = \omega^{D/24} q^{-D/12} f^{-D}(q^2). \quad \text{(III.9)}$$

This yields

$$T_1{}^{\text{pl}} = g^4 \int_0^1 dq q^{-\alpha_P(0)-1} f^{-24}(q^2) \int_{0=\Theta_1 \leqq \Theta_2 \leqq \Theta_3 \leqq \Theta_4 \leqq \pi} \prod_{i=2}^{4} d\Theta_i \prod_{1 \leqq i < j \leqq 4} \bar{\psi}(\Theta_j - \Theta_i)^{-p_i p_j} \quad \text{(III.10)}$$

and

$$T_i{}^{\text{npl}} = g^4 \left(\frac{\pi}{2}\right)^t \int_0^1 dq q^{-\alpha_P(t)-1} f^{-24}(q^2) \int \prod_{i=2}^{4} d\Theta_i \prod_{1 \leqq i < j \leqq 4} [\bar{\psi}(\Theta_j - \Theta_i) \quad \text{or} \quad \bar{\psi}_T(\Theta_j - \Theta_i)]^{-p_i p_j} \quad \text{(III.11)}$$

with the integration region

$$0 = \Theta_1 \leqq \Theta_4 \leqq \pi; \qquad \Theta_3 \leqq \Theta_2 \leqq \pi + \Theta_3, \qquad 0 \leqq \Theta_3 \leqq \pi.$$

The functions $\bar{\psi}(\Theta)$, $\bar{\psi}_T(\Theta)$ of eq. (III.10, 11) are defined by $\psi(\Theta, q)/(-\ln \omega/\pi)$ or $\psi_T(\Theta, q)/(-1/2\, q^{-1/4} \ln \omega)$, respectively, and $\alpha_P(t) = (D - E)/12 + t/4 = 2 + t/4$ where $t = (p_1 + p_4)^2 = (p_2 + p_3)^2$.

Let us stay for a moment at the nonplanar loop formula (III.11). Due to the behaviour of the integrand near $q = 0$ $(\omega = 1)$ there arises a new Regge trajectory $\alpha_P(t)$ in the t-channel which has a slope of half the usual one and a doubled intercept $\alpha_P(0) = 2\alpha(0) = 2$. If one further introduces isospin quantum numbers by the Chan-Paton method, the diagram of fig. 16b gets a factor $\text{Tr}(\lambda_1\lambda_4) \cdot \text{Tr}(\lambda_2\lambda_3) = \text{Tr}(\lambda_4\lambda_0)\,\text{Tr}(\lambda_0\lambda_2\lambda_3)$ so that the new singularity exchanged in the t-channel has the quantum numbers of the vacuum.

---

[8]) We recall that the partition function arises as a result of the calculation of the operator trace III.2. It is given as $[f(\omega)]^{-D} = \text{Tr}\, \omega^H$.

Finally, the diagram contains cuts in the s-channel which is dual to the vacuum t-channel as can be seen by investigating the integral (III.11) in the vicinity of $q = 1(\omega = 0)$ [62]. Thus it is tempting to associate the new singularity with the Pomeron singularity which is here dual to a cut-background in accord with the HARARI-FREUND conjecture [63]. The pole nature of the new singularity depends crucially on the fact that we used $D = D_{\text{crit.}} = 26$, $E = 2$, $F = 0$ and $a = 1$ as has been first pointed out by LOVELACE [64]. If these conditions are not fulfilled, there appear additional ($\log q$)-factors in the integrand of eq. (III.11) converting the Pomeron pole into a cut which gives rise to logarithmic branch point singularities at $t_n = -(D - E)/3 + 4n$, $(n = 0, 1, 2, \ldots)$ in the t-channel. As these branch points do not correspond to normal thresholds (they are independent of the intercept $a = -2m^2$ of the input trajectory), they violate perturbative unitarity. Thus, we see the surprising result that the nonplanar orientable one-loop diagram forces us to work in the critical dimension of space time of the ghost-free DRM with Virasoro gauges ($a = 1$).[9])
After this discussion of the nonplanar loop graph let us return to the one-loop planar diagram and ask what kind of singular behaviour arises there for $q \to 0(\omega \to 1)$.
As follows from eq. (III.10), the integral representation is now divergent because of the factor $q^{-\alpha_p(0)-1} = q^{-3}$. Due to eq. (III.9) this divergence is connected with the exponential divergence of the partition function $f^{-24}(\omega) \xrightarrow[\omega \to 1]{} \exp 4\pi^2/(1 - \omega)$ which reflects the large degeneracy of levels contributing in the trace calculation (comp. footnote p. 66). The question of the regularization of the planar loop integral will be postponed to sect. III.4. where the renormalization problem of the dual loop amplitudes will be discussed.
For the following it will be useful to have operatorial expressions for the planar and nonplanar self-energy graphs shown in fig. 17a + b which will serve later on as building blocks for the construction of multi-loop amplitudes.[10])

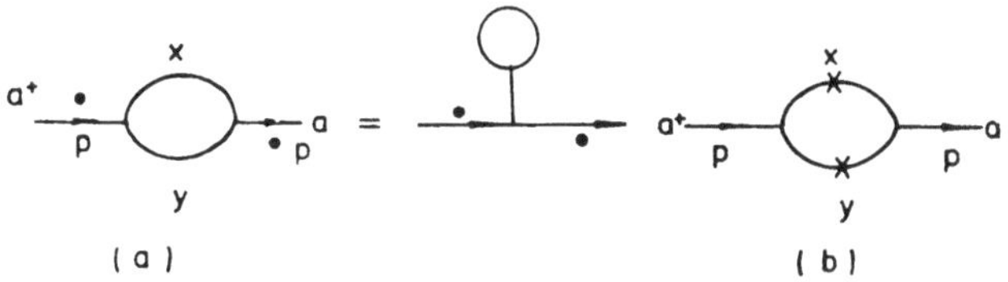

Fig 17 The planar (a) and nonplanar orientable one-loop self-energy operator (b)

The corresponding operator forms have been obtained by sewing three-Reggeon vertices with nontwisted or twisted propagators [66—68] or by factorizing the one-loop amplitude [69].
This gives (see eq. (II.40) for notations)

$$\Sigma^{(T)}(a^+, a; p) = g^2 \int d\tau^{(T)} \exp \left\{ p^2 \varphi^{(T)} + (p|\, F^{(l)}\, |a^+) + \frac{1}{2}\, (a^+|\, E^{(l)}\, |a^+) \right\}$$

$$\times :\exp\left(-(a^+|\, X - 1\, |a)\right): \exp \left\{ (p|\, F^{(r)}\, |a) + \frac{1}{2}\, (a|\, E^{(r)}\, |a) \right\}. \qquad \text{(III.12)}$$

---

[9]) In order to be consistent with another requirement of (nonperturbative) unitarity — the Froissart theorem [65] — one should have $\alpha_{p(\text{ren})}(0) \leq 1$, where the subscript denotes the inclusion of renormalization effects (see Sect. III.3.1.3. If there is still the relation $\alpha_{p(\text{ren})}(0) = 2\alpha_{\text{ren}}(0)$ we should wait for a more physical DRM with $\alpha_{(\text{ren})}(0) = 1/2$ possessing nevertheless enough gauge conditions.
[10]) The dots appearing in the planar self-energy operator describe the orientation of the external lines. They arise from the symmetrical vertices used and are remnants of the kind of factorization shown in fig. 6. Sewing together planar self-energy parts with a planar (twisted) propagator gives then a twisted (nontwisted) internal line.

Here, $d\tau^{(T)}$ is a measure function ($\omega = xy$)

$$d\tau^{(T)} = d\omega \, \frac{dx}{x} \left(-\frac{2\pi}{\ln \omega}\right)^{D/2} \frac{(1-\omega)^{-F,(G)}}{\omega^{\alpha+1}} \, f^{E}(\omega) \, f^{-D}(\omega) \, h^{(T)}(x, \omega) \tag{III.13}$$

where the functions $\varphi, h, \varphi^T, h^T$ of the planar and nonplanar self-energy operator are given by

$$\varphi = \frac{1}{2} \ln \frac{\psi^2(x, \omega)}{(1-x)\left(1-\frac{\omega}{x}\right)}, \qquad h(x, \omega) = \left[(1-x)\left(1-\frac{\omega}{x}\right)\right]^{-1}$$

$$\varphi^T = \ln \frac{\psi_T(x, \omega)}{1-\omega}, \qquad h^T(x, \omega) = 1. \tag{III.14}$$

The ω and x variables have to be integrated within $[0, 1]$ and $[\omega, 1]$, respectively. Furthermore, the 'left' (l) and 'right' (r) matrices appearing in the exponent of eq. (III.12) are essentially derivatives of the logarithm of elliptic functions. For explicit representations of these matrices we refer the reader to refs. [*67*, *68*].

For further applications we quote also the Jacobi transformed version of the nonplanar self-energy operator Σ^T obtained by factorizing the amplitude (III.11) [*69*]

$$\Sigma^T(a^+, a; p) = \frac{q^2}{\pi^2} \, 2^{-p^2} \int_0^1 dq q^{-\alpha_p(p^2)-1} f^{-24}(q^2) \int_0^\pi d\Theta [\overline{\psi}_T(\Theta)]^{p^2} F(a, a^+; q, \Theta). \tag{III.15}$$

Here, the operatorial expression $F(a, a^+; q, \Theta)$ has the same exponential structure as in eq. (III.12) with analogous E, F and X matrices (see eqs. (B. 2—4), appendix).

Concluding this section we remark that we omitted the discussion of the nonplanar nonorientable loop graphs from the very beginning. The reason for this is that they contain unwanted exotic intermediate states if one includes $SU(N)$ quantum numbers by the Chan-Paton method.

III.2.2. Generalization to the NS- and NSR-model

For completeness, we quote also briefly the results obtained in the NS-(NSR-)model for planar and nonplanar one-loop meson amplitudes with internal meson (fermion) propagators. As the relevant propagators and vertices factorize into separate a- and b-(d) mode parts, the only modifications relative to the GVM concern the evaluation of traces of Fermi-like b-(d-)mode operators. This yields ($D = D_{\text{crit}} = 10$) [*70*] ($\omega = x_1 \cdot x_2 \cdot x_3 \cdot x_4$)

$$T^{\text{pl}}_{1,\text{NS}} = \left(\frac{1}{2} g_{\text{NS}}\right)^4 (2\pi)^{-5} \int_0^1 \left(\prod_{i=1}^4 dx_i\right) \omega^{-3/2} \ln^{-5} \omega \left[\frac{\varphi_+(\omega)}{f(\omega)}\right]^8 \prod_{1 \le i < j \le 4} \psi(x_{ij})^{-p_i p_j}$$

$$\times \sum_{P \in P'} (-1)^P \prod_{j=1}^2 p_{i_{2j-1}} p_{i_{2j}} X^+(x_{i_{2j-1}, i_{2j}}, \omega) \tag{III.16}$$

where the sum runs over all permutations $P \in P'$ that group the external particles in pairs, each pairing being counted only once regardless of the ordering in each pair. The new functions X^+ and φ^+ are defined according to

$$X^+(x, \omega) = \sum_{n=1}^{\infty} \frac{x^{n-1/2} + \left(\frac{\omega}{x}\right)^{n-1/2}}{1 + \omega^{n-1/2}} = \frac{i}{2}\, \Theta_2(0 \mid \tau)\, \Theta_4(0 \mid \tau)\, \frac{\Theta_3(\nu \mid \tau)}{\Theta_1(\nu \mid \tau)}$$

$$\varphi^+(\omega) = \prod_{n=1}^{\infty} (1 + \omega^{n-1/2}) \tag{III.17}$$

where

$$\nu = \frac{\ln x}{2\pi i} \qquad \tau = \frac{\ln \omega}{2\pi i}$$

and Θ_1, Θ_2, Θ_3, Θ_4 are Jacobi's theta functions. The planar amplitude of the NSR-model (planar-fermion loop) follows from eq. (III.16) if one formally replaces $\omega^{-3/2}$ by ω^{-1}, g_{NS}^4 by $(2g_{\mathrm{R}})^4$ and the functions $\varphi_+(\omega)$, $X^+(x, \omega)$ by $\varphi_0(\omega)$, $2X_F{}^+(x, \omega)$ which are defined by

$$X_F{}^+(x, \omega) = \frac{1}{2} + \sum_{n=1}^{\infty} \frac{x^n + \left(\frac{\omega}{x}\right)^n}{1 + \omega^n} = \frac{i}{2}\, \Theta_3(0 \mid \tau)\, \Theta_4(0 \mid \tau) \cdot \frac{\Theta_2(\nu \mid \tau)}{\Theta_1(\nu \mid \tau)}$$

$$\varphi_0(\omega) = \prod_{n=1}^{\infty} (1 + \omega^n). \tag{III.18}$$

As usual there is an overall minus sign for a closed fermion loop. For the nonplanar case we quote only the results obtained after Jacobi transformation [70, 71] ($z_1 = y_1 = 0$)

$$T_{1,\mathrm{NS}}^{\mathrm{npl}} = \frac{2^{p^2}}{\pi^2} \int_0^1 dq\, q^{-\alpha_p(p^2)} \left[\frac{\varphi^+(q^2)}{f(q^2)}\right]^8 \cdot \int_0^\pi d\Theta \int_0^1 dy \int_0^1 dz \left\{\frac{1}{\pi}\, \overline{\psi}(\pi z)\right\}^{-p_1p_2} \left\{\frac{1}{\pi}\, \overline{\psi}(\pi y)\right\}^{-p_3p_4}$$

$$\times \left(\prod_{i=1}^{2} \prod_{j=3}^{4} \{\overline{\psi}_T(\Theta + \pi(y_i + z_i))\}^{-p_ip_j}\right) \{\det [p_ip_j \cdot \pi \overline{\chi}_T{}^+(i, j)]\}^{1/2}. \tag{III.19}$$

Here, $(-\pi/\ln \omega)\, \overline{\chi}^+(i, j)$ are the Jacobi transforms of the X^+ functions (III.17) and of a similar defined function $X_T{}^+(x, \omega)$

$$X_T{}^+(x, \omega) = \sum_{n=1}^{\infty} \frac{(-x)^{n-1/2} - \left(-\frac{\omega}{x}\right)^{n-1/2}}{1 + \omega^{n-1/2}} = \frac{i}{2}\, \Theta_2(0 \mid \tau)\, \Theta_4(0 \mid \tau)\, \frac{\Theta_1(\nu \mid \tau)}{\Theta_2(\nu \mid \tau)} \tag{III.20a}$$

where the arguments (i, j) are the same as those of the functions raised to the power $(-p_ip_j)$.

Finally, the nonplanar loop formula of the NSR-model [72] can be formally obtained from eq. (III.19) by replacing $\varphi_+(q^2) \to \varphi_0(q^2)$ and $\overline{\chi}_T \to \overline{\chi}_{F,T}^+$ where

$$\overline{\chi}_{F,T}^+ = -\Theta_2(0 \mid \tau)\, \Theta_3(0 \mid \tau)\, \frac{\Theta_2(\nu \mid \tau)}{\Theta_3(\nu \mid \tau)} \left(-\frac{\pi}{\ln \omega}\right)^{-1}. \tag{III.20b}$$

The '+' sign at the X- and φ-functions distinguishes them from X^- and φ^- functions arising in the case of a 'negative G-parity Pomeron' not considered here. The elimination of spurious states yields again a power $[f(\omega)/\varphi_+{}^0(\omega)]^2$ leading to an effective power $D_{\mathrm{eff}} = D_{\mathrm{crit}} - 2 = 8$ used in the above formulae.

III.2.3. The dual Pomeron singularity

Let us go back once more to the nonplanar loop expressions (III.11, 15) ($D = D_{\text{crit}} = 26$, $a = 1$). There arises naturally the question whether the spectrum and the coupling of the new 'Pomeronic' particles can be described in terms of operatorial expressions, too. That this can indeed be done has been shown by CREMMER and SCHERK [69] and CLAVELLI and SHAPIRO [71]. It turns out that the Pomeron sector of the DRM may be characterized by two sets of mutually commuting operators c_n, $\bar{c}_n$ possessing analogous commutators as in eq. (II.18).
One introduces, in particular, a Pomeron propagator $D_P(p^2)$

$$D_P(p^2) \cdot P = \int_0^1 \frac{dq}{q} f^2(q^2)\, q^{L_0^c + L_0^{\bar{c}} - 2} \oint \frac{dz}{2\pi i z} z^{L_0^c - L_0^{\bar{c}}} \tag{III.21}$$

(P is the projection operator defined by the contour integral in (III.21) ensuring that for any state $|\Phi\rangle$ in the pomeron sector $L_0^c|\Phi\rangle = L_0^{\bar{c}}|\Phi\rangle$). After introducing a Reggeon-Pomeron transition vertex $V_P(a^+, c, \bar{c})$ (see ref. [69] for definition), the nonplanar self-energy operator Σ^T can be refactorized in the form of a tree graph operator, as shown in fig. 18[11])

$$\Sigma^T(a^+, a; p) = \langle 0_{c,\bar{c}}| V_P(a^+, c, \bar{c})\, |0_a\rangle\, D_P(p^2)\, P \cdot \langle 0_a|\, V_P(a, c^+, \bar{c}^+)\, |0_{c,\bar{c}}\rangle \tag{III.22}$$

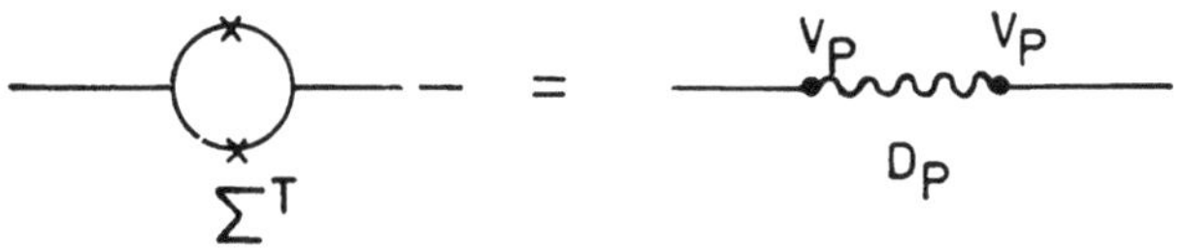

Fig. 18. Factorization of the nonplanar self-energy operator Σ^T

There is the surprising result that the Pomeron sector coincides with the nonplanar Shapiro-Virasoro model [14]. Using gauge conditions one can furthermore prove that the Pomeron sector is again ghost-free and that it is saturated by the transverse states of the Shapiro-Virasoro model. The latter fact explains naturally the occurrence of the factor $f^2(q^2)$ in the Pomeron propagator.
Concluding this section, we mention that the selfenergy operator $\Sigma^T(a^+, a; p)$ may also be used to construct more complicated one-loop diagrams containing external particles of arbitrary (orbital) spin. For example, a dual Pomeron coupling to four excited particles with occupation numbers $\{l_i\}$ ($i = 1, 2, 3, 4$) is defined by the expectation value

$$T_{\{l_1\},\cdots,\{l_4\}} = \langle O_{1,2,3,4}|\, \langle O_a|\, DV(a_1, a_2, a; p_1, p_2, -p)\, \Sigma^T(a^+, a; p)$$
$$\times\, DV(a_2, a_4, a^+; p_3, p_4, p)\, |O_a\rangle\, |\{l_1\}\rangle_1\, |\{l_2\}\rangle_2\, |\{l_3\}\rangle_3\, |\{l_4\}\rangle_4 \tag{III.23}$$

where DV is a three-Reggeon vertex with a propagator attached. Amplitudes of such a type are necessary to describe diffraction dissociation. They provide also informations on the helicity conservation properties of the dual Pomeron and on the Pomeron form

[11]) Because of eq. (III.22) the notation 'self-energy' operator seems to be now somewhat misleading. We shall nevertheless keep this notation also in the following.

factor (comp. sect. III.3.1.4.) [*73*]. Furthermore, dual amplitudes have been constructed for the scattering of any number of external Pomerons and Reggeons [*74*], too.

III.2.4. Multi-loop amplitudes

III.2.4.1. Iteratively constructed amplitudes

We are now ready to generate multi-loop operatorial expressions by inserting the planar or nonplanar selfenergy operators $\Sigma^{(T)}(a^+, a; p)$ into a propagator and iterating them.[12] Thus we get the following operator equations for the 'complete' propagators of 'ϱ'- and 'f'-trajectories (comp. fig. 19a–b)[13]

$$\Delta_\varrho = \frac{1}{L_0 - 1} + \frac{1}{L_0 - 1}\, g^2\Sigma(a^+, a; p)\, \Delta_\varrho \tag{III.24}$$

$$\Delta_f = \frac{1}{L_0 - 1} + \frac{1}{L_0 - 1}\, [g^2\Sigma(a^+, a; p) + g^2\Sigma^T(a^+, a; p)]\, \Delta_f. \tag{III.25}$$

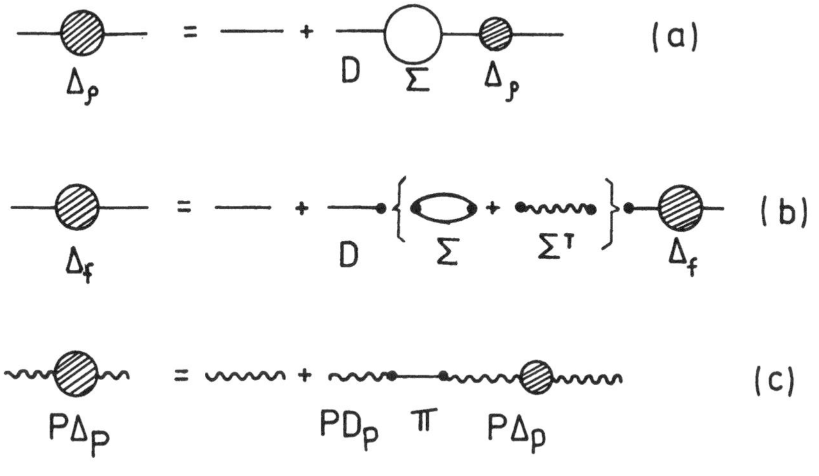

Fig. 19. Iteration of the 'ρ', 'f'- and Pomeron propagators (a–c)

In eqs. (III.24, 25) use has been made of the fact that Σ^T couples only to the f-trajectory. A similar equation may be derived also for the complete Pomeron propagator Δ_P if one introduces a Pomeron transition operator π by

$$\pi(c^+, \bar{c}^+, c, \bar{c}; p) = \langle O_a|\, V_P(a, c^+, \bar{c}^+)\, \frac{1}{L_0 - 1}\, V_P(a^+, c, \bar{c})\, |O_a\rangle \tag{III.26}$$

so that (see fig. 19c)

$$P \cdot \Delta_P = PD_P + PD_P g^2 \pi P \Delta_P. \tag{III.27}$$

---

[12]) As the problem of eliminating spurious states from multiloop amplitudes is not yet solved, we use here for simplicity non-projected propagators in sewing self-energy parts.

[13]) Moreover, one can take also into account signature by using a signatured propagator $[1 \pm (-1)^H] \times (L_0 - 1)^{-1}$. The simplified twisted propagator $(-1)/(L_0-1)^{-1}$ has been proposed in ref. [*60*] for loop calculations with a physical state projector τ.

The equations (III.24, 25, 27) may be formally solved yielding

$$\Delta_\rho = [L_0 - 1 - g^2\Sigma(a^+ a; p)]^{-1} \equiv \int_0^1 dx x^{L_0 - 2 - g^2\Sigma(a^+, a; p)} \tag{III.28}$$

$$\Delta_f = \left[L_0 - 1 - g^2\left(\Sigma(a^+, a; p) + \Sigma^T(a^+, a; p)\right)\right]^{-1} \tag{III.29}$$

$$P\Delta_P = [(PD_P)^{-1} - g^2\pi(c^+, \bar{c}^+, c, \bar{c}; p)]^{-1}. \tag{III.30}$$

From eqs. (III.28—30) we see that the self-energy insertions modify the 'bare' propagators quite similarly as in conventional field theory. In particular, one would expect to obtain also new trajectory functions with cuts in their respective Mandelstam variable and, possibly, with renormalized values of the trajectory parameters a and α'.
Let us show that there arise indeed new complex trajectories. For this purpose, we take the matrix element of the propagator (III.28) between occupation number states $|\{l^{\text{lead}}\}\rangle$ belongig to the parent trajectory ($j = l = \sum_{\mu_1=0}^{D-1} l_{1,\mu_1}$, see eq. (II.46)), neglecting, for simplicity, nondiagonal terms corresponding to parent-daughter transitions. Then

$$\langle\{l^{\text{lead}}\} |\Delta_\varrho| \{l^{\text{lead}}\}\rangle = \frac{1}{l - \alpha(p^2) - g^2\Sigma(p^2, l) + O(g^4)} \tag{III.31}$$

where $\Sigma(p^2, l)$ is defined by the 'parent contribution' to the matrix element of $\Sigma(a^+, a; p)$

$$\langle\{l^{\text{lead}}\}| \Sigma(a^+, a; p) |\{l^{\text{lead}}\}\rangle = \Sigma(p^2, l) + \text{daughter contributions} \tag{III.32}$$

and

$$\Sigma(p^2, l) = \int d\tau(x, \omega) \left[\frac{\psi(x, \omega)^2}{(1 - x)\left(1 - \frac{\omega}{x}\right)}\right]^{p^2/2} (X_{11})^l. \tag{III.33}$$

The matrix element X_{11} reads

$$X_{11} = (1 - x)(1 - y) \left(\frac{2}{\ln \omega}\right)^2 \frac{\partial^2}{\partial \left(\frac{\pi \ln x}{\ln \omega}\right)^2} \ln \psi(x, \omega) \tag{III.34}$$

and the measure $d\tau(x, \omega)$ was defined in eqs. (III.13, 14). Taking in the same way the matrix element of Δ_f gives besides (III.33) an additional nonplanar trajectory correction

$$\Sigma^T(p^2, l) = \int d\tau^T(x, \omega) \left[\frac{\psi_T(x, \omega)}{1 - \omega}\right]^{p^2} (X_{11}^T) \tag{III.35}$$

where now

$$X_{11}^T = -(1 - \omega)^2 \frac{\partial^2}{\partial\lambda\partial\mu} \ln \psi_T(x\lambda\mu)\Big|_{\lambda=\mu=1} \tag{III.36}$$

and $d\tau^T(x, \omega)$ is again given by eqs. (III.13, 14).
The new (corrected) trajectory functions $\alpha_\rho^{\text{new}}(t)$, $\alpha_f^{\text{new}}(t)$ are then defined by the vanishing of the denominators of the matrix elements of Δ_ρ and Δ_f

$$\alpha_\rho^{\text{new}}(t) = \alpha(t) + g^2\Sigma\left(\alpha_\rho^{\text{new}}(t), t\right) + O(g^4) = \alpha(t) + g^2\Sigma\left(\alpha(t), t\right) + O(g_4) \tag{III.37}$$

$$\alpha_f^{\text{new}}(t) = \alpha(t) + g^2\left(\Sigma\left(\alpha_f^{\text{new}}(t), t\right) + \Sigma^T\left(\alpha_f^{\text{new}}(t), t\right)\right) + O(g^4). \tag{III.38}$$

In order to obtain for example a four-point function one has to compute the matrix elements of the complete propagators between the 'coherent states' $\langle O| V(p_1)$ and $V(p_2) |O\rangle$.
Unfortunately, we were not able to obtain a compact expression. Of such an amplitude by performing the necessary complicated algebra. Instead, it is however possible to construct explicit expressions for M-loop amplitudes. Studying the high energy behaviour of these amplitudes enables us again to get informations on the trajectory functions of the complete amplitude [75].
The M-loop amplitudes may easily be written down in terms of M-loop self-energy operators $\Sigma_M^{(T)}(a^+, a; p)$, defined by the recursion formulae[14]) [68, 75]

$$\Sigma_M^{(T)} = \Sigma_{M-1}^{(T)} \frac{1}{L_0 - 1} \Sigma^{(T)} \; (\Sigma_1^{(T)} \equiv \Sigma^{(T)}) \tag{III.39}$$

where the operators $\Sigma_M^{(T)}$ are again given by a quadratic exponential form of the type (III.12) and all relevant quantities are now depending on the index M. Eq. (III.39) implies, in particular, that there exists a multiplication law for quadratic exponential forms.
In appendix B we quote the M-loop matrices E_M, X_M, F_M and the M-loop integration measure $d\tau_M$ in the form of an iteration scheme. (If the twisted propagator of ref. [60] (footnote [13])) had been used in eq. (III.39) one obtains the same iteration scheme with the replacement $z_{M-1} \to -z_{M-1}$.
The four-point M-loop amplitude now easily follows by attaching external vertices and propagators (with Chan variables u, v) to $\Sigma_M^{(T)}(a^+, a; p)$ and taking the corresponding matrix element. We obtain[15])

$$\begin{aligned} T_M(s, t) = {} & g^{2(M+1)} \int d\varrho_M(x_i, \omega_i, z_j, u, v) \\ & \times \exp \left\{ \frac{s}{2} (1| \{u\} X_M\{u\} |1) - \frac{t}{2} \left[(1| F_M^{(l)}\{u\} |1) + (1| F_M^{(r)}\{u\} |1) - 2\varphi_M \right] \right. \\ & \left. + 2(1| \{u\} X_M\{u\} |1) - (1| \{u\} E_M^{(l)}\{u\} |1) - (1| \{u\} E_M^{(r)}\{u\} |1) \right\} \end{aligned} \tag{III.40}$$

where we have used the notation (II.40), $\{z\}_{mn} = z^n \delta_{mn} (n, m, = 1, 2, \ldots)$ and $d\varrho_M(x_i, \omega_i, z_j, u, v) = du\, dv (uv)^{-\alpha(t)-1} d\tau_M(x_i, \omega_i, z_j)$. Note that there are $3M + 1$ integration variables in eq. (III.40) ($M - 1$ Chan variables z_j of propagators connecting one-loop selfenergy parts, M one-loop variables x_i, ω_i and 2 Chan variables u, v of 'boundary' propagators). The Chan variables u, v, z_j and the loop variables ω_i have to be integrated over $[0, 1]$, whereas the loop variables x_i run over the interval $[\omega_i, 1]$.
Concluding this section, we mention that the above multiloop amplitudes are not free of spurious states. In order to avoid these one has to recalculate the multi-loop amplitudes for $D = 26$ space-time dimension using the on-shell physical state projector of Brink and Olive [61]. Waiting for a solution of this rather complicated task, we nevertheless expect that such a projection does not essentially change the general structure of the above iteration scheme. In particular, the D appearing in the power of the multi-loop partition function $[\det (\Phi_M)^{-1}]^{-D/2}$ (comp. eq. B.6) should be replaced by an effective power $D - 2$ as there propagates in this case only the spectrum of the 24-dimensional transverse (DDF) oscillators. Such a '$D - 2$' rule would be consistent with the results of the one-loop case [60].

---

14) For simplicity, we restrict us here to a separate iteration of planar and nonplanar operators.
15) The nonplanar M-loop amplitude could also be constructed by defining a $(M - 1)$-loop Pomeron transition operator Π_{M-1} which should be sandwiched between tree states ending with a Pomeron propagator.

III.2.4.2. Loop expressions as Abelian integrals

III.2.4.2.a. The analogue model

In this section we shall sketch an approach to the multiloop problem developped by LOVELACE [76] and ALESSANDRINI, AMATI [77, 78] that provides an intersting analogue model picture of the integrand of the M-loop amplitude: if one identifies the quark lines of the duality diagram of the M-loop amplitude (see fig. 20) with the boundaries of a two-dimensional conductor with holes cut out and the momenta of the external particles with the currents of electrodes, the integrand of the M-loop amplitude may be interpreted as the exponential of the heat generated by the currents in the conductor.

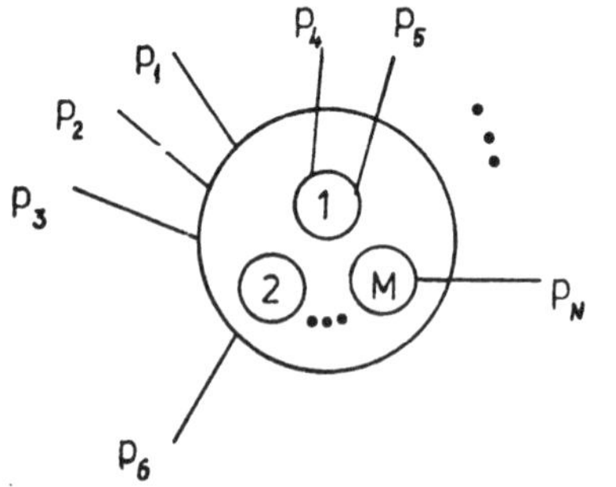

Fig. 20. The analogue model picture of the N-point M-loop amplitude

This generalizes an idea first pointed out by FAIRLIE and NIELSEN [16] for the dual tree and one-loop amplitudes[16]). To determine the M-loop integrand, one has thus to solve a boundary value problem of the second kind with the derivatives of the potential (the currents) given on the boundaries. After having solved this boundary value problem by the Neumann function $N(z_i, z_j)$ the momentum-dependent part of the integrand of the M-loop amplitude should be given by

$$\exp\left[-\frac{\alpha'}{2}\sum_{i\neq j} p_i p_j N(z_i, z_j)\right]. \tag{III.41}$$

The Neumann functions involved here are those discovered by BURNSIDE [79] some time ago already. They are formulated most conveniently in terms of the so-called third Abelian integrals.

In the following we shall collect some general formulae appearing in the theory of functionals on Riemann surfaces [76–80]. These formulae will then be compared with the multi-loop expressions derived from the operator formalism. Finally, we shall apply some of these ideas to the study of a nonplanar orientable two-loop amplitude which represents an iteration of the bare dual Pomeron [81]. We are going to follow closely the representations contained in refs. [77, 78].

The construction of Green or Neumann functions of an open Riemann surface B proceeds as follows: First one constructs a new closed Riemann surface associated to B, its 'double' D. Then the Abelian integrals on the double D are defined allowing transparent expressions for the Green or Neumann function on the original surface B. Let us con-

[16]) In this picture the Jacobi transformed variable q of eqs. (III.8, 10, 11) is just the radius of the hole. A planar diagram is characterized by the fact that all particles are attached to the outer boundary of fig. 20 whereas in the nonplanar case particles are attached also to the hole boundaries.

sider, for example, the duality diagram B of fig. 20. Taking an identical mirror image $\bar{B}$ of B and sewing together their boundaries gives as their double a sphere with M handles. This surface is shown in fig. 21, together with a set of 'vertical' and 'horizontal' cycles $C_{2\mu-1}$, $C_{2\mu}$ ($\mu = 1, 2, \ldots, M$) running around the handles.

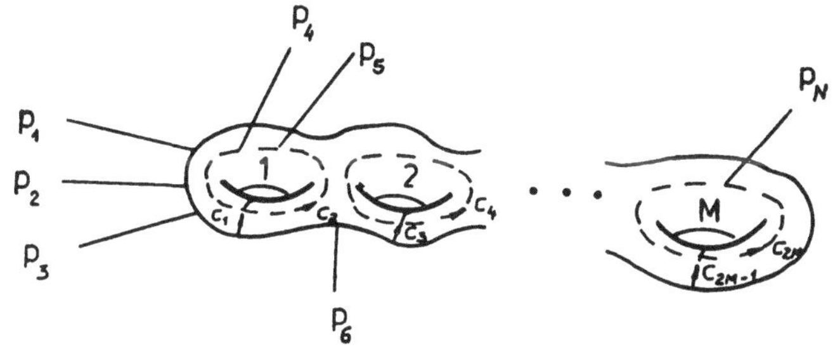

Fig. 21. Sphere with M handles, attached to the M-loop amplitude with 'horizontal' and 'vertical' cycles $C_{2\mu-1}$, $C_{2\mu}$

The particles are now attached to the 'horizontal' cycles and to the cycle which encircles all handles. Indeed, cutting the surface of fig. 21 along the 'horizontal' cycles and the cycle that goes around all the M handles gives again the open surfaces B and $\bar{B}$ of fig. 20. Cutting the double along the horizontal 'boundary' lines only and bending and stretching it yields a surface which is topologically equivalent to a complex plane with $2M$ holes the boundaries of which are pairwise identified. Similarly, one can cut the double along its 'vertical' cycles mapping it topologically onto another complex plane. In fig. 22 we show a particular mapping where the M pairs of equally sized circles $C_{2\mu-1}$, $C'_{2\mu-1}$ originally sewn together, lie on the KN-unit circle.

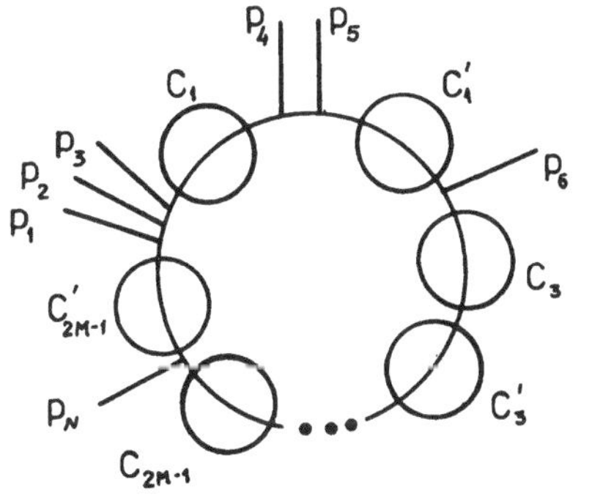

Fig. 22. The (Koba-Nielsen-)complex plane with circles $C_{2\mu-1}$, $C'_{2\mu-1}$ obtained from cutting the sphere with M handles along 'vertical' cycles

As we see, internal 'boundary' lines running around the handles are now mapped into arcs between the pairs of circles $C_{2\mu-1}$, $C'_{2\mu-1}$ whereas the 'boundary' line encircling all handles is mapped onto the arcs between neighbouring circles $C_{2\mu-1}$, $C'_{2\mu+1}$ ($\mu = 1, 2, \ldots, M$). In order to parametrize these complex planes one has yet to explain, how to 'identify' points on neighbouring circles $C_{2\mu-1}$, $C_{2\mu-1}$ ($\mu = 1, 2, \ldots, M$) to obtain handles. Mathematically this identification will be defined as a mapping of the circle $C_{2\mu-1}$ onto $C_{2\eta-1}$ generated by a projective transformation $T_\mu(z)$. As circles $C_{2\mu-1}$, $C'_{2\mu-1}$ one can choose the 'isometric' circles of $T_\mu(z)$ and of its inverse $T_\mu^{-1}(z)$. Let us write

$$T_\mu(z) = \frac{a_\mu z + b_\mu}{c_\mu z + d_\mu}; \qquad T_\mu^{-1}(z) = \frac{-d_\mu z + b_\mu}{c_\mu z - a_\mu} \qquad (a_\mu d_\mu - b_\mu c_\mu = 1) \qquad \text{(III.42)}$$

or, equivalently, in terms of the multiplier K_μ and the invariant points ξ_μ, η_μ [82]

$$\frac{T_\mu(z) - \xi_\mu}{T_\mu(z) - \eta_\mu} = K_\mu \frac{z - \xi_\mu}{z - \eta_\mu}. \tag{III.43}$$

The isometric circles $C_{2\mu-1}$, $C'_{2\mu-1}$ may then be defined by $|c_\mu z + d_\mu| = 1$, $|c_\mu z - a_\mu| = 1$. Their radius ε_μ and their centers M_μ, M_μ' are given by

$$M_\mu' = \frac{a_\mu}{c_\mu} = \frac{\xi_\mu - K_\mu \eta_\mu}{1 - K_\mu}, \qquad M_\mu = -\frac{d_\mu}{c_\mu} = \frac{\eta_\mu - K_\mu \lambda_\mu}{1 - K_\mu},$$
$$\varepsilon_\mu = \frac{1}{|c_\mu|} = \frac{\sqrt{K_\mu}\, |\xi_\mu - \eta_\mu|}{1 - K_\mu}. \tag{III.44}$$

The complex plane with M pairs of identified holes is thus characterized by $3M$ parameters K_μ, ξ_μ, η_μ — the so-called Riemann moduli. The i) first, ii) second and iii) third Abelian integrals on the double D are now by definition functions whose differentials are i) regular at every point of D, ii) regular except for a double pole, iii) regular except for two single poles with opposite residues. It can be shown that a sphere with M handles has M independent first Abelian differentials $d\varphi_\nu(\xi)$. In particular, one can choose a set with normalized periods around the vertical cycles $C_{2\eta-1}$

$$P_\eta(d\varphi_\nu) = \oint_{C_{2\mu-1}} d\varphi_\nu = \delta_{\mu\nu} \qquad (\mu, \nu = 1, \dots, M). \tag{III.45}$$

The periods of the first Abelian differentials along the horizontal cycles $C_{2\mu}$ define the period matrix $A_{\mu\nu}$

$$A_{\mu\nu} = 2\pi i \oint_{C_{2\mu}} d\varphi_\nu. \tag{III.46}$$

In the following we need also the third Abelian integrals in 'real' or 'complex normalization' $\Omega_{ab}(\xi)$ and $\omega_{ab}(\xi)$. Their differentials are regular on D except for poles at $\xi = a$ and b with residues ± 1 and are normalized according to

$$\mathrm{Re} \oint_{\substack{C_{2\mu-1} \\ \text{or } C_{2\mu}}} d\Omega_{ab}(\xi) = 0; \qquad \oint_{C_{2\mu-1}} d\omega_{ab}(\xi) = 0. \tag{III.47}$$

In particular, one can show that $\Omega_{ab}(\xi)$ and $\omega_{ab}(\xi)$ differ only by a linear combination of first Abelian integrals which do not modify their singularities.

$$\Omega_{ab}(\xi) = \omega_{ab}(\xi) - \sum_{\mu,\nu=1}^{M} \varphi_\eta(\xi)\, (A^{-1})_{\mu\nu}\, \mathrm{Re}\, [\varphi_\nu(a) - \varphi_\nu(b)]. \tag{III.48}$$

As shown by Schiffer and Spencer [80], the Neumann function of the original Riemann surface B may then be expressed in terms of the third Abelian integrals $\Omega_{ab}(\xi)$ in the following way:

$$N(\xi, \eta, \eta_0) = N(\xi, \eta) - N(\xi, \eta_0) = \frac{1}{2} [\Omega_{\eta\eta_0}(\xi) + \Omega_{\bar\eta\bar\eta_0}(\bar\xi) + \Omega_{\eta\eta_0}(\bar\xi) + \Omega_{\bar\eta\bar\eta_0}(\xi)]. \tag{III.49}$$

Here ξ, η, η_0 are points of B and $\bar{\eta}, \bar{\eta}_0, \bar{\xi}$ their respective 'conjugate' points (mirror images) on the double D[17]). If the points ξ, η lie on the boundary ($\xi = \bar{\xi}$ etc.) one obtains a particular simple expression ($\xi, \eta \rightarrow z_i, z_j$)

$$N(z_i, z_j) - N(z_i, z_0) = \Omega_{z_l z_0}(z_j) + \Omega_{z_l z_0}(\bar{z}_j) = \mathrm{Re}\, \Omega_{z_l z_0}(z_j). \tag{III.50}$$

The main task is to find the single-valued meromorphic differential $d\Omega_{\eta\eta_0}(\xi)$ and by integration, $\Omega_{\eta\eta_0}(\xi)$. Let us now consider a single-valued function $\varphi(\xi)$ on the sphere with M handles shown in fig. 21. If one goes around any handle (cycle) it should reproduce its value. Using further the special Riemann surface of fig. 22 it has to be periodic both in going around the circles $C_{2\mu-1}, C'_{2\mu-1}$ and in going from one circle $C_{2\mu-1}$ to its neighbour $C'_{2\mu-1}$ (we recall that the latter operation is just mediated by the projective transformation $T_\mu(z)$). Thus, we have

$$\varphi(T_\mu(z)) = \varphi(z) \qquad (\mu = 1, \ldots, M) \tag{III.51a}$$

and, more generally,

$$\varphi(T(z)) = \varphi(z) \tag{III.51b}$$

where $T(z)$ is any element of the 'automorphic group' defined by all the products of arbitrary positive and negative powers of the generator transformations T_μ ($\mu = 1, 2, \ldots, M$). Moreover, the region of the complex plane, exterior to all the isometric circles $C_{2\mu-1}, C'_{2\mu-1}$ is the 'fundamental region' of the automorphic group.
The single-valued functions with the periodicity property (III.51) are called automorphic functions. Thus, the doubly-periodic elliptic functions appearing in the one-loop case with one generator $T_1(z)$ (comp. eqs (III.6, 7) are the simplest automorphic functions. The automorphic functions can be constructed using the Poincare theta-series

$$\vartheta(z, a) = \sum_\alpha \frac{1}{(c_\alpha z + d_\alpha)^2} \frac{1}{T_\alpha(z) - a} \tag{III.52}$$

where the summation extends to all the elements of the automorphic group. We quote only the final results for the first and third Abelian integral and the period matrix

$$\varphi_\mu(z_i) - \varphi_\mu(z_j) = \sum_{P\alpha \in Q_\mu} \log \left(\frac{T_\alpha(z_i) - \xi_\mu}{T_\alpha(z_i) - \eta_\mu} \cdot \frac{T_\alpha(z_j) - \eta_\mu}{T_\alpha(z_j) - \xi_\mu} \right) \tag{III.53}$$

$$\omega(z_i, z_j; z_k, z_l) \equiv \omega_{z_k z_l}(z_i) - \omega_{z_k z_l}(z_j) = \sum_\alpha \log \left(\frac{z_i - T_\alpha(z_k)}{z_i - T_\alpha(z_l)} \cdot \frac{z_j - T_\alpha(z_l)}{z_j - T_\alpha(z_k)} \right) \tag{III.54}$$

$$A_{\mu\nu} = \delta_{\mu\nu} \log \omega_\mu + \sum_{T\alpha \in Q_{\mu\nu}} \log \left(\frac{\xi_\mu - T_\alpha(\xi_\nu)}{\xi_\mu - T_\alpha(\eta_\nu)} \cdot \frac{\eta_\mu - T_\alpha(\eta_\nu)}{\eta_\mu - T_\alpha(\xi_\nu)} \right) \tag{III.55}$$

where the summation region $Q_{\mu\nu}$ is the set of group elements which have *not* a power of T_μ on the left and $Q_{\mu\nu}$ is the set of group elements which have not a power of T_μ on the left and not a power of T_ν on the right. For $\mu = \nu$ the identity is excluded.

[17]) This just corresponds to the 'method of images' or 'method of inverse radii' in potential theory. For $M = 0$, $\Omega_{\eta\eta_0}(z) = \ln (z - \eta_0/z - \eta)$ is the potential generated by two unit charges of opposite strength in the point z. The double of a disc with radius R is the whole complex z-plane and the conjugate points are given by $\bar{z} = R^2/\bar{z}$.

III.2.4.2.b. The multi-loop formula

In the foregoing section we have given only a summary of properties of invariant functions on Riemann surfaces needed to formulate Fairlie's and Nielsen's conjecture (III.41) for the general case of a N-point amplitude with M loops. A proof of this conjecture as well as an explicit expression of the measure of the M-loop amplitude missing in the analogue model should yet be delivered directly in terms of the operator formalism. This proof has been first carried out by LOVELACE [76]. One can start from the $N + 2M$ Reggeon vertex [19] (comp. sect. II.1.9.) and sew together the M pairs of Reggeons a_μ, $\tilde{a}_\mu$ joined by (untwisted or twisted) propagators Δ_μ by means of a functional integration. The M-loop amplitude with N external scalar particles then reads [78]

$$A_{M,N} = \int_0^1 \prod_{\mu\in[M]} \frac{dx_\mu(-1)^a}{x_\mu^{1+a}(1-x_\mu)^{1-a}} \int \left(\prod_{i=1}^{2M+N} \frac{dz_i}{d^3V_{abc}} \frac{1}{z_{i+1}-z_i} \left(\frac{z_{i+1}-\beta_{i-1}}{z_i-\beta_{i-1}}\right)^a \right)$$

$$\times \prod_{i\in[N]} e^{(\mu^2/2)\varphi_i} \frac{\exp\left[-\frac{1}{2}\sum_{i,j\in[N]} p_ip_jN(z_i,z_0;z_j,z_0')\right]}{(\det A)^2\,[\det(1-\Sigma)]^2}. \tag{III.56}$$

Here we used $\varphi_i = (U^i)_{00} + (V^i)_{00}$ and $N(z_i, z_0; z_j, z_0')$ is expressed by the Neumann function of a sphere with M handles if $i \neq j$ and by the regular part $N - \log|1 - z_i/z_j|$ if $i = j$. For z_i, z_j on the unit circle, we have

$$\left.\begin{aligned} N(z_i,z_0;z_j,z_0') &\equiv [N(z_i,z_j)-N(z_i,z_0)] - [N(z_i,z_0')-N(z_i,z_0)] \\ &= \operatorname{Re}\Omega_{z_iz_0}(z_j) - \operatorname{Re}\Omega_{z_iz_0}(z_0') \\ &= \operatorname{Re}\Omega_{z_jz_0'}(z_i) - \operatorname{Re}\Omega_{z_jz_0'}(z_0) \end{aligned}\right\} \tag{III.57}$$

where

$$N(z_i,z_j) = (z_i|\frac{1}{1-\Sigma}|z_j) = \sum_{\mu,\nu=1}^{M} (z_i|\frac{1}{1-\Sigma}(T_\mu^{-1}-1)|\tilde{z}_\mu)$$

$$\times (A^{-1})_{\mu\nu}(\tilde{z}_\nu|(T_\nu-1)\frac{1}{1-\Sigma}|z_j) \tag{III.58}$$

$$\left(|z)_n = \frac{z^n}{\sqrt{n}}, \quad (z|_n = \frac{z^{*n}}{\sqrt{n}} \qquad (n = 1, 2, 3, \ldots)\right)$$

and

$$\Sigma = \sum_{\mu=1}^{M} (T_\mu + T_\mu^{-1}). \tag{III.59}$$

Here, the T_μ are definite dimensional representations of certain projective transformations being the generators of the automorphic group. They arise naturally by the sewing of the μ-th Reggeon pair in the N-Reggeon vertex and are given in the sense of eq. (II.68) by

$$T_\mu = V_\mu D_\mu U_\mu = \begin{bmatrix} \infty & 0 & 1 \\ \tilde{\beta}_{\mu-1} & \tilde{z}_r & \tilde{z}_{\mu+1} \end{bmatrix} \begin{bmatrix} 0 & \infty & 1 \\ 0 & \infty & x_\mu \end{bmatrix} \begin{bmatrix} \beta_{\mu-1} & z_\mu & z_{\mu-1} \\ 0 & \infty & 1 \end{bmatrix}. \tag{III.60}$$

In eq. (III.60) we have used eqs. (II.69, 73).

For the sewing with a twisted propagator one obtains an analogous equation where one has now to take, however, eq. (II.70). Moreover, using $[1 - \Sigma]^{-1} = \sum_\alpha T_\alpha$ and

$$\sum_\alpha (z_i|\, T_\alpha \,|z_j) \approx -\sum_\alpha \log |z_i - T_\alpha(z_j)| \approx \mathrm{Re}\, \omega_{z_i z_0}(z_j) \tag{III.61}$$

we see that the first term in eq. (III.58) is just the third Abelian integral in 'complex normalization' (comp. with eq. (III.55)). The second term in eq. (III.58) arises from the loop momentum integrations. It is given by a product of the real parts of first Abelian integrals which can be rewritten in the form of eq. (III.53) with the period matrix A given by (comp. with eq. (III.55))

$$A_{\mu\nu} = (\tilde{z}_\nu|(T_\nu - 1) \frac{1}{1 - \Sigma} (T_\mu^+ - 1)|\tilde{z}_\mu). \tag{III.62}$$

For a specification of the integration region in eq. (III.56) the reader is referred to ref. [*78*].

Concluding this section we remark that it is difficult to compare the loop expressions (III.40) and (III.56) directly as different computational methods have been used in deriving these amplitudes. In the iterative approach, the loop momentum integrations were performed first to produce the self-energy operators. Then these operators were sewn together. In the other approch these two steps were performed in reversed order. The above formulae have subsequently been generalized to the NS-model [*83*].

III.2.4.2.c. An example: The nonplanar orientable two-loop graph

Let us now apply the multi-loop formula (III.56) to the nonplanar orientable two-loop graph shown in fig. 23 [*81*].

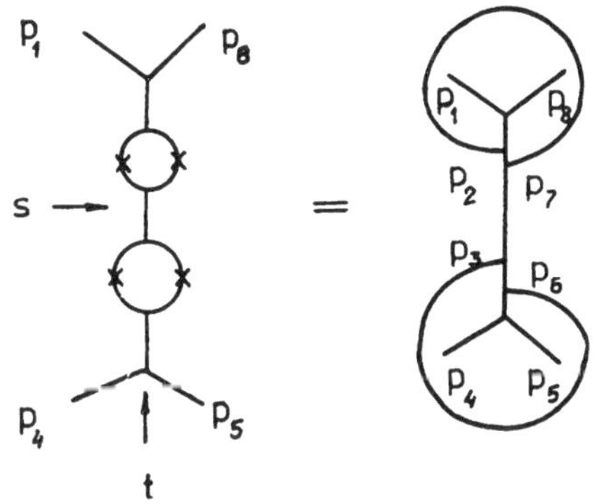

Fig. 23. The non-planar two-loop amplitude as obtained from the eight-Reggeon vertex

This graph is interesting because it contains an iterated Pomeron in the t-channel. Moreover, it is one of the simplest non-trivial generalizations of the one-loop case confronting us with the task of handling automorphic functions. In fig. 23 we have shown the duality transformation which brings the diagram in the form of an 8-Reggeon vertex with two pairs of sewn legs called '3' and '7'. The two generators of the associated sphere with two handles (complex plane with two pairs of isometric circles cut out) are then given using eq. (III.60) by

$$T_3 = \begin{bmatrix} \beta_3 & z_3 & z_4 & \beta_2 \\ z_5 & z_7 & z_6 & \beta_6 \end{bmatrix}, \quad T_7 = \begin{bmatrix} \beta_7 & z_7 & z_8 & \beta_6 \\ z_1 & z_3 & z_2 & \beta_2 \end{bmatrix} \tag{III.63}$$

where in this notation any one of the columns is determined by the other three.

Furthermore the momentum dependent part of the integrand yields the following s- and t-dependence of the integrand

$$\exp\left\{\frac{s}{2}\, N(z_8, z_1; z_4, z_5) + \frac{t}{2}\, N(z_1, z_4; z_5, z_8)\right\}. \tag{III.64}$$

As we have shown the operator formalism provides a definite parametrisation of the sphere with handles. In particular, the isometric circles of the generators (III.60) correspond just to the vertical cycles on this surface. There arises the question how one obtains from this representation the 'orthogonal' parametrisation defined by cutting the surface along 'horizontal' cycles. This parametrisation corresponds just to the analogue model description and will be shown to be particular useful for determining the asymptotical behaviour of the graph of fig. 23. The transformation from the 'vertical' to the 'horizontal' parametrisation is nothing else than the Jacobi transformation already discussed in the one-loop case which has to be defined now for the multi-loop amplitude, too.

Let us denote the Jacobi transformed generators of the above two-loop graph by P_7 and P_3, by ϱ_μ, σ_μ, $\omega_\mu (\mu = 3,7)$ their respective invariant points and multipliers and by y_i the Jacobi transform of the original KN-variable z_i. In fig. 24 we have illustrated the mapping under the Jacobi transformation. In particular, points z (with $|z| = 1$) between the isometric circles (e.g. z_1, z_8, z_4, z_5) have two images in the fundamental region (y-plane without the interior of the isometric circles). To be definite we choose y_1, y_8 to lie on the circle around σ_7 and y_4, y_5 on that around σ_3.

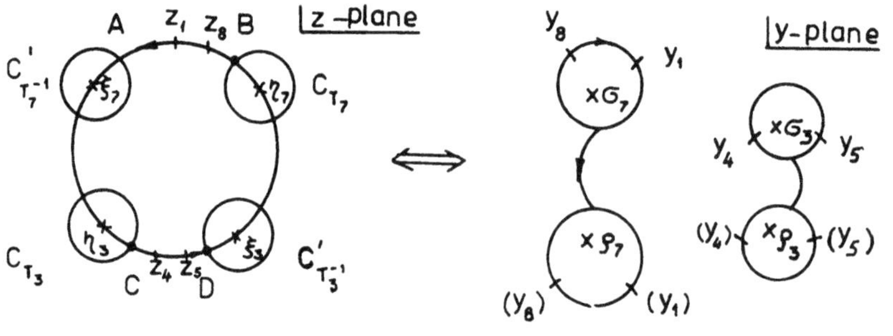

Fig. 24. The mapping of the original K – N complex plane under the Jacobi transformation $y(z)$. The arcs AB, CD are mapped on the isometric circles of P_7 and P_7^{-1} or of P_3 and P_3^{-1}, respectively The isometric circles of T_3, T_3^{-1} and T_7, T_7^{-1} are mapped on curves connecting identified points on the isometric circles of P_i, P_i^{-1} ($i = 3,7$)

As the Neumann function and the third Abelian integrals are invariant functions on their respective Riemann surface, we may write

$$N(z_8, z_1; z_4, z_5) = \hat{N}(y_8, y_1; y_4, y_5) \quad \text{etc.} \tag{III.65}$$

where $\hat{N}$ is the Jacobi transformed Neumann function. Now, the Abelian integrals and the period matrix have the same representations (III.53—55) but, of course, transformed variables y_i, σ_μ, ϱ_μ, ω_μ have to be used. Because of eq. (III.65) the momentum dependent part (III.64) of the integrand of the two-loop amplitude is trivially transformed. In order to investigate also the behaviour of the integration measure of the loop integrand a more detailed study of the Jacobi transformation is necessary. For this purpose, let us mention that the uniqueness theorem of first Abelian integrals leads to the following implicit definition of the Jacobi transformation $y = y(z)$

$$\varphi_\mu(z) - \varphi_\mu(z_0) = 2\pi i \sum_{\nu=3,7} (\hat{A}^{-1})_{\mu\nu} \left(\hat{\varphi}_\nu(y) - \hat{\varphi}_\nu(y_0)\right). \tag{III.66}$$

Furthermore, integrating eq. (III.66) along the 'horizontal' cycles and using the normalization of the transformed functions $\hat{\varphi}_\nu(y)$ as well as the definition of the period matrix yields the following relation between the period matrix A and the Jacobi transformed period matrix $\hat{A}$

$$4\pi^2 A^{-1} = \hat{A}. \tag{III.67}$$

By eq. (III.67) one obtains a functional relation between the original and the Jacobi transformed multipliers and invariant points. In particular, we shall use the transformations (III.66, 67) in the case when the handles of the sphere (the holes of the plane) shrink to zero (comp. sect. III.3.2.). Let us now discuss the asymptotic behaviour of the multi-loop amplitudes and their corrections to the bare Regge trajectories of the tree amplitudes.

III.3. High energy behaviour

III.3.1. Asymptotic behaviour of planar and nonplanar orientable multi-loop graphs

In this section we shall investigate the high energy limit of the M-loop amplitude (III.40) for both the planar and nonplanar cases. As the iteration of self-energy parts is mediated by Regge poles as shown in fig. 25.

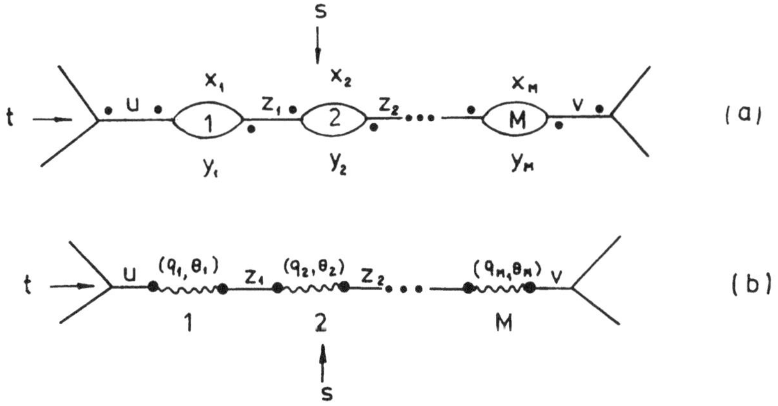

Fig. 25. Iterated planar (a) and nonplanar (b) M-loop amplitudes with integration variables attached.

We expect to obtain a Regge multipole behaviour. In particular, it should be possible to sum up for high energies the leading contributions to all orders of the dual coupling constant g to obtain a simple Regge behaviour of the total amplitude with new corrected trajectories $\alpha_{\varrho,f}^{\text{new}}(t)$, $\alpha_{\text{p}}^{\text{new}}(t)$ (comp. eq. (III.37, 38). In ref. [75] we have proved that the multi-loop integrand possesses indeed the required factorization properties necessary to perform the summation. Let us quote the main results. For this, we first rewrite the amplitude (III.40) as follows ($u = z_0$, $v = z_M$)

planar case:

$$T_M^{\text{pl}}(s, t) = g^{2(M+1)} \int \prod_{i=0}^{M} dz_i z_i^{-\alpha(t)-1} \int d\mu(z_0, \ldots, z_M) \exp\left\{-\frac{s}{2}\, U_s^{(M)} - \frac{t}{2}\, U_t^{(M)}\right\} \tag{III.68}$$

nonplanar case:

$$T_M^{\mathrm{npl}}(s, t) = g^{2(M+1)} \int \prod_{i=1}^{M} dq_i q_i^{-\alpha_p(t)-1} \int \prod_{j=0}^{M} dz_j z_j^{-\alpha(t)-1}$$
$$\times \int d\mu(z_0, \ldots, z_M; q_1, \ldots, q_M) \exp\left\{-\frac{s}{2} V_s^{(M)} - \frac{t}{2} V_t^{(M)}\right\}. \quad \text{(III.69)}$$

For convenience, the remaining loop variables have not been made explicit here but are contained in the integration measure $d\mu$ that can be read off together with the functions $U^{(M)}$, $V^{(M)}$ from the original representation (III.40). Moreover, eq. (III.69) is obtained by iterating the Jacobi transformed operator (III.15).

The limit $|s| \to \infty$ to be studied will be performed parallel to the imaginary axis so that we have to investigate the asymptotic behaviour of multiple Fourier integrals.[18]) According to a generalized multidimensional 'method of stationary phase' [*84*], the Regge behaviour of the amplitudes (III.68, 69) arises from those 'critical' end-points where $U_S^{(M)} = 0$ or $V_S^{(M)} = 0$ on the boundary of the integration region. These functions are then stationary in any direction on the tangent plane to the boundary at that point. Next, one expands the integrand near the critical points, integrates and adds the asymptotic series contributed by every critical point.

Let us first quote, as an example the leading contribution to eq. (III.68) corresponding to the simultaneous factorization of all the intermediate Regge pole of a given kind. (It may be obtained by expanding the integrand around the corner $z_0 = z_1 = \cdots = z_M = 0$ of the integration region; comp. Appendix A). Using now the fact that the measure $d\mu$ and the function $U_s^{(M)}$ turn out to factorize when expanded around the respective corner of the integration region yields a typical integral of the form

$$g^{2(M+1)} \int \left(\prod_{j=1}^{M} d\mu^0(j)\right) \int_0^1 \prod_{i=0}^{M} dz_i z_i^{-\alpha(t)-1} \exp\left\{-\frac{s}{2} (-z_0 z_1 \cdots z_M) \prod_{j=1}^{M} U_s^0(j)\right\}$$
$$\underset{|s|\to\infty}{\sim} \frac{g^2}{M!} \left(\frac{\partial}{\partial l}\right)^M \left\{\left(-\frac{s}{2}\right)^l \Gamma(-l) \left[g^2 \int d\mu^0 (U_s^0)^l\right]^M\right\}\Bigg|_{l=\alpha(t)}. \quad \text{(III.70)}$$

Eq. (III.70) exhibits the expected Regge multipole asymptotics $(-s/2)^{\alpha(t)} \ln^M s$. Let us apply this formula now in some detail.

III.3.1.1. Trajectory corrections in the planar case

Using eq. (III.40) and the iteration equations (B.1) we obtain

$$-U_s^M = (1| \{z_0\} X_M\{z_M\} |1) \sim (z_0 z_1 \cdots z_M) \left[\prod_{j=1}^{M} (X(\omega_i, x_i))_{11}\right] \equiv (z_0 z_1 \cdots z_M) \prod_{j=1}^{M} U_s^0 \quad . \quad \text{(III.71)}$$

where $(X)_{11} \leqq 0$ is the (11)-matrix element of the one-loop matrix X defined in eq. (III.34). The asymptotic behaviour of the planar multiloop amplitude is then given by eq. (III.70) where the expression in square bracket is nothing else than the parent matrix element $\Sigma(t, l)$ (III.33) of the planar self-energy operator $\Sigma(a^+, a, p)$ already com-

[18]) This restriction is necessary, at least, for the nonplanar amplitudes as there the integral representation converges only inside a strip $-t < s < 4m^2$. The question of the analytic continuation of the final results into the whole s-plane as well as the existence of saddle points inside the integration region has been discussed extensively in the one-loop case [*62*].

puted by NEVEU and SCHERK [85] in the one-loop case. The analytical and asymptotic properties of the planar self-energy correction have been investigates in ref. [85]. The behaviour (III.70, 71) has also been found independently by KARPF and LIEHL [86] using domain variational methods for Abelian integrals. Taking $a = 1$, $D = D_{\text{crit}} = 26$, $E = 2$, $F = 0$ (Comp. eq. (III.13)) and performing the Jacobi transformation, (III.8) yields another useful representation $(\alpha_p(0) = 2)$

$$\Sigma(t, l) = (-1)^l \int_0^1 dq q^{-\alpha_p(0)-1} f^{-24}(q^2) \int_0^\pi d\Theta (\bar{\psi}(t))^t \left[-\frac{\partial^2 \ln \bar{\psi}(\Theta)}{\partial \Theta^2} \right]^l. \tag{III.72}$$

The integrand of the self-energy 'correction' $\Sigma(t, l)$ contains the same singular behaviour near $q = 0$ as the one-loop amplitude (III.10). The resulting divergence of the integral is closely related to the Pomeron singularity — a fact which is of crucial importance for defining a counterterm and a renormalization scheme (comp. sect. III.4.).

Finally, if we sum up the leading contributions (III.70) over M the following asymptotic behaviour of the total amplitude results[19])

$$T(s, t) = \sum_{M=0}^{\infty} T_M(s, t) \underset{|s| \to \infty}{\sim} g^2 Z(t) \, \Gamma\big(-\alpha_{\text{new}}(t)\big) \left(-\frac{s}{2} \right)^{\alpha_{\text{new}}(t)} \tag{III.73}$$

where the new trajectory function $\alpha_{\text{new}}(t)$ is defined by a functional (bootstrap) equation identical to eq. (III.37) and

$$Z(t) = \left[1 - g^2 \frac{\partial}{\partial l} \Sigma(t, l)|_{l = \alpha_{\text{new}}(t)} + 0(g^4) \right]^{-1}. \tag{III.74}$$

We thus find the anticipated result that the new Regge behaviour is determined by the corrected trajectory function of the complete propagator introduced in sect. III.2.4.1. The function $Z(t)$ may be interpreted as a Reggeon 'Z-factor' defined by the residuum of the matrix element of the complete propagator at the new pole position $l = \alpha_{\text{new}}(t)$.

III.3.1.2. Residue and higher order trajectory corrections

The result (III.73) takes into account only trajectory corrections of order $0(g^2)$ leading to a corrected 'Reggeon' propagator' $Z(t)\, \Gamma(-\alpha_{\text{new}})\, (-\alpha' s)^{\alpha_{\text{new}}}$ but leaving the residue g^2 unaltered. Such a contribution arises when all intermediate parent Regge poles were simultaneously factorized out. Let us now discuss for the example of the two-loop graph how there appear residue and higher order trajectory corrections, too. We expect that different dual contributions of a given diagram contribute to the asymptotic behaviour as shown in fig. 26.

Here, the wiggly lines indicate 'asymptotic' (factorized) Regge poles. Furthermore, in order to avoid multiple counting both the Regge residues (form factors) and higher order self-energy parts should be irreducible with respect to the parent trajectory. This means that the dashed lines have to include some subtraction in order not to give rise to a dual Regge pole in the t-channel. It can be shown that such residue and higher order self-energy corrections indeed appear if one takes into account the contributions of the different critical boundaries of the integrand during the course of the asymptotic computations (see appendix A). Using eqs. (A.6, B.1) the $O(g^4)$ self-energy correction of

---

19) There are some slight differences to the respective formulae of ref. [75] because there the twisted projected propagator (II.66) has been used for the sewing of the self-energy parts.

fig. 26c for example, turns out to be

$$g^2 \frac{\partial}{\partial l} \left\{ \left(-\frac{s}{2}\right)^l \Gamma(-l) \, [g^4 \int d\tau(1) \int d\tau(2) \int dz z^{-\alpha(t)-1}(1 - O_z) \right.$$

$$\left. \times \left(X(1) \{z\} \tilde{\Phi}_2 X(2)\right)_n^l (\det \Phi_2)^{D-E/2} \, e^{[\varphi(1)+\varphi(2)]t} \right\} \Bigg|_{l=\alpha(t)} . \qquad \text{(III.75)}$$

Here, $d\tau(i)$, $X(i)$ ($i = 1, 2$) are the one-loop integration measure and matrices defined by eqs. (III.12–14), $(1 - O_z)$ is a subtraction prescription (comp. eq. (A.5)). The two-loop partition function $[\det \Phi_2]^{D-E/2}$ as well as the functions $\tilde{\Phi}_2$, φ_2 are defined in appendix B. As expected, eq. (III.75) leads to a double Regge pole behaviour $\sim s^{\alpha(t)} \cdot \ln s$.

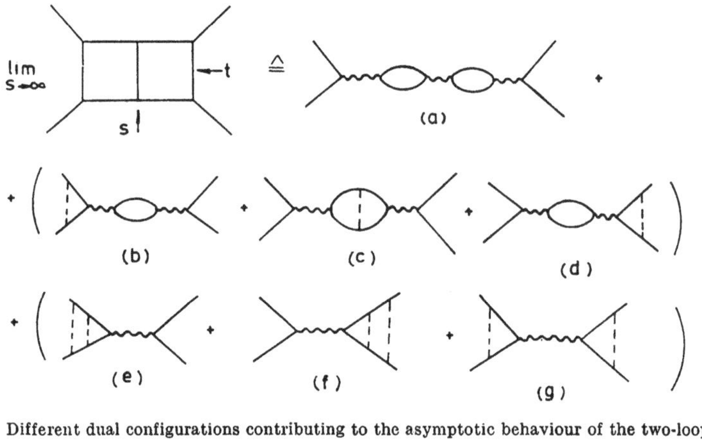

Fig. 26. Different dual configurations contributing to the asymptotic behaviour of the two-loop graph

In the same way we may compute the contributions of the remaining asymptotic-diagrams, too, leading also to residue corrections. Thus, we obtain for fig. 26g

$$\left(-\frac{s}{2}\right)^{\alpha(t)} \Gamma(-\alpha(t)) \left[g^3 \int d\tau(1) \int_0^1 dz z^{-\alpha(t)-1}(1 - O_1) \left(\{(1| \{z\} X(1)\}_1\right)^{\alpha(t)} \right.$$

$$\left. \times \exp\left\{ -\frac{t}{2} [(1| F(1) \{z\} |1) - 2\varphi(1)] - (1| \{z\} E(1) \{z\} |1) \right\} \right] \times [\text{'2'}] \qquad \text{(III.76)}$$

where ['2'] is the transposed expression [...] depending now on the one-loop variables '2'. Eq. (III.76) yields a residue correction of order $O(g^3)$. Summing up all such contributions of higher loop graphs, too, yields again a Regge behaviour as in eq. (III.73) where now $Z(t)$ and $\alpha_{\text{new}}(t)$ contain the $O(g^4)$ trajectory correction defined by eq. (III.75) and where g^2 is replaced by a residue function given by

$$\beta(t) = g^2\left(1 + g^2\beta^{(1)}(t) + O(g^4)\right)^2 \qquad \text{(III.77)}$$

with $\beta^{(1)}(t)$ defined by eq. (III.76).
As we see, the above kind of working resembles very much to that used for studying the high energy behaviour of ladder graphs in conventional quantum field theory [87]. There, the links of the ladder have to be 'contracted' in all possible ways in order to obtain both trajectory and residue corrections of the out-put Regge trajectory.

III.3.1.3. Corrections to the Pomeron trajectory

The corrections to the bare Pomeron trajectory from the nonplanar multi-loop amplitudes (III.69) can now be computated analogously. For simplicity, we give only the asymptotic result arising from the corner $q_1 = q_2 = \cdots = q_M = 0$ of the integration region. Using the expansion formulae (B. 2—5) we obtain

$$V_s^{(M)} \sim -(q_1 q_2 \cdots q_M)\,(1|\,\{z_0\}\left[\prod_{i=1}^{M-1} X^0(\Theta_i)\,\{z_i\}\,\tilde{\Phi}^0(z_i)\right] X^0(\Theta_M)\,\{z_M\}\,|1)$$

$$\equiv -(q_1 q_2 \cdots q_M)\,(1|\,\{z_0\}\,X_M{}^0\{z_M\}\,|1). \tag{III.78}$$

Eq. (III.78) may be further factorized by performing a suitable change of variables $\Theta_i \to \varrho_i$ [75] (Compare footnote[19]))

$$(1|\,\{z_0\}\,X_M{}^0\,\{z_M\}\,|1) = 8 \sin \pi z_0 \sin \pi z_M \cos \varrho_M \prod_{j=1}^{M-1} C(\varrho_j, z_j) \tag{III.79}$$

where

$$C(\varrho, z) = |e^{i\varrho}(k|\,\{z\}\tilde{\Phi}^0\,|k) + e^{-i\varrho}(k|\,\{z\}\,\tilde{\Phi}^0|k)|$$

$$(k|_n = {}_n|k) = \frac{\sqrt{n}}{n!}\,(2\pi i)^n. \tag{III.80}$$

Now one can perform the q_i-integrations by means of formula (III.70) ($z_i \to q_i$) and then integrate over ϱ_M using

$$\int_0^{2\pi} d\varrho_M(-\cos \varrho_M)^{\alpha_P} = (1 + e^{-i\pi\alpha_P})\,B\left(\frac{\alpha_P + 1}{2}, \frac{1}{2}\right). \tag{III.81}$$

This yields the final result ($D = 26$, $E = 2$, $a = 1$)

$$T_M(s, t) \underset{|s|\to\infty}{\sim} \frac{\pi^t}{(M-1)!}\left(\frac{\partial}{\partial l}\right)^{M-1} \left\{\left[\left(-\frac{s}{2}\right)^l + \left(\frac{s}{2}\right)^l\right] 2^l \right.$$

$$\left.\times\, B\left(\frac{l+1}{2}, \frac{1}{2}\right) \Gamma(-l)\, f_P{}^2(t, l)\,[g^2\,\pi(t, l)]^{M-1}\right\}\Bigg|_{l=\alpha_P(t)} \tag{III.82}$$

where $f_P(t, \alpha_P(t))$ is a Pomeron 'form factor' defined by

$$f_P(t, \alpha_P(t)) = 2^{\alpha_P(t) - \alpha(t)}\, 2\pi g \int_0^1 dz\,[\sin \pi z]^{\alpha_P(t) - \alpha(t) - 1} = 2^{\alpha_P(t) - \alpha(t) + 1} \cdot g \cdot B\left(\frac{\alpha_P(t) - \alpha(t)}{2}, \frac{1}{2}\right) \tag{III.83}$$

and $\pi(t, \alpha_P(t))$ a Pomeron-Reggeon transition part given by

$$\pi(t, \alpha_P(t)) = \frac{2^{-t+1}}{\pi^2} \int_0^{\pi} d\varrho \int dz\, \frac{z^{-\alpha(t)-1}}{[\det(\Phi_2{}^0)^{-1}]^{12}}\,[F(z)]^t\,[C(\varrho, z)]^{\alpha_P(t)}. \tag{III.84}$$

The functions $C(\varrho, z)$, $F(z)$ are defined in eqs. (III.80) and (B.5). Eq. (III.82) reproduces for $M = 1$ just the one-loop results found in ref. [62]. As shown by eqs. (III.82, 83) the dual Pomeron trajectory has positive signature as well as an 'f-dominated' residue. Indeed, from eq. (III.83) follows that the poles of $f_P(t, \alpha_P(t))$ are located at positions $\alpha_P(t) - \alpha(t) + 2m = 0$ $(m = 0, 1, 2, \ldots)$. The factorization properties of the asymptotic M-loop amplitude (III.82) allow again to perform the sum over M which yields a Regge behaved amplitude with a corrected Pomeron trajectory

$$\alpha_P^{\text{new}}(t) = \alpha_P(t) + g^2 \pi\left(t, \alpha_P^{\text{new}}(t)\right) + O(g^4). \tag{III.85}$$

We mention that the z-integral in eq. (III.84) is divergent due to the singular behaviour of the two-loop partition function [det $(\Phi_2{}^0)^{-1}]^{-12}$ at $z = 1/4$ [75]. Therefore, the region of the z-integration has to be restricted to $0 \leqq z \leqq 1/4$.

III.3.1.4. Corrections to the ordinary trajectory from the nonplanar loop; spin modifications

Besides the corner $q_i = 0$ $(i = 1, \ldots, M)$ yielding an asymptotic Pomeron multipole behaviour $\sim |s/2|^{\alpha_P} \ln^{M-1}|s/2|$ there are other critical boundaries of the integrand of the nonplanar M-loop amplitude. The give rise to a behaviour $\sim |s/2|^{\alpha} \ln^M|s/2|$, where $\alpha = \alpha(t)$ is the coupling 'f'-trajectory. In particular, one easily sees that $z_i = 0$ $(0 \leqq i \leqq M)$ or $z_0 = 1$ and $z_M = 1$ are also critical boundaries. Restricting ourselves to the contribution of all simultaneously factorized intermediate ordinary Regge poles we have to consider only the four 'corners' $(i = 1, \ldots, M-1)$

$$(z_0; z_i; z_M) = (O; O; 0), \quad (1; 0; 1), \quad (0; 0; 1) \quad \text{and} \quad (1; 0; 0).$$

Adding their contributions gives a signaturized asymptotic behaviour [75]

$$T_M{}^{\text{npl}}(s, t) \underset{|s|\to\infty}{\sim} \frac{g^2}{M!}\left(\frac{\partial}{\partial l}\right)^M \left\{\left[\left(-\frac{s}{2}\right)^l + \left(\frac{s}{2}\right)^l\right] \Gamma(-l)\, [g^2 \Sigma^T(t, l)]^M\right\}\bigg|_{l=\alpha(t)} \tag{III.86}$$

with the nonplanar trajectory correction

$$\Sigma^T(t, l) = \frac{2^{-t}}{\pi^2} \int_0^1 dq q^{-\alpha_p(t)-1} f^{-24}(q^2) \int_0^\pi d\Theta [\overline{\psi}_T(\Theta)]^l \left[-\pi^2 \frac{\partial^2}{\partial \Theta^2} \ln \overline{\psi}_T(\Theta)\right]^l. \tag{III.87}$$

Eq. (III.87) is nothing but the self-energy correction (III.35) written in terms of Jacobi transformed variables. It develops poles at $l = \alpha_P(t) - 2m$ $(m = 0, 1, \ldots)$ due to the q-integration. Again we can sum up to obtain a corrected ('f') trajectory $\alpha_f{}^{\text{new}}(t) = \alpha(t) + g^2\Sigma^T(t, \alpha_f{}^{\text{new}}(t)) + O(g^4)$. In the case of simultaneous iteration of planar and nonplanar self-energy insertions with quantum numbers attached their would arise a different correction scheme for the 'ϱ'- and 'f'-trajectories (comp. eq. (III.37, 38). Thus the unitarity corrections break the exchange degeneracy of the bare trajectories appearing in the tree amplitudes.

Finally, we want to discuss shortly some modifications of the asymptotic behaviour (III.82) due to the inclusion of spin or b-mode degrees of freedom for the external particle of the conventional or the NSR-model, respectively. Restricting in the first case to a process $j_1 + O \to j_2 + 0$, for simplicity, we obtain an asymptotic behaviour of the type (III.82) with, of course, some tensor structure and the following Pomeron form factor for spin couplings [73, 56]

$$f_{(m,n)}^{j_1,j_2}\left(t, \alpha_P(t)\right) = 2^{\alpha_P(t)-\alpha(t)+1} g \cdot B\left(\frac{\alpha_P(t) - \alpha(t) + m}{2}, \frac{j_1 + j_2 + 1 - n}{2}\right) \tag{III.88}$$

(m, n are suitably restricted numbers). Inserting the identity $\alpha_P(t) - \alpha(t) = -\alpha_P(t) + 3$ into the beta function (III.88) one sees that the spin of the external particles shifts the form factor zeros located at $\alpha_P(t) \geqq 4$ in th spinless case to $\alpha_P(t) \geqq (4 + j_1 + j_2)$ or to $\alpha_P(t) \geqq \min(4 + j_1 + j_2, 4 + j_3 + j_4)$ if also $j_3 \neq 0$ and $j_4 \neq 0$. Thus, for higher spins of the external particles higher particle poles on the Pomeron trajectory (compensated in the scalar case by form factor zeros) are able to couple to the amplitude. This generalizes a result obtained in ref. [71] for scalar-vector scattering. Moreover, we have found that the dual Pomeron does not exhibit any helicity conservation property except for s-channel forward scattering. There exists a nonvanishing leading contribution to the reaction $j_1 + j_3 \to j_2 + j_4$ only if $\sum_{i=1}^{4} j_i =$ even. Similar results have been found in ref. [88].

The multi-loop formula (III.40) may be directly generalized to the (NS—NSR-)model. For this one defines, for example, by factorizing eq. (III.19) the nonplanar self-energy operator $\Sigma^T(a^+, a; c^+, c; p)$ of the NS-model ($c = b$) or NSR-model ($c = d$), respectively. The asymptotic behaviour of the M-loop amplitude with a'positive G-parity' Pomeron looks again like eq. (III.82) where now the integrand of the function $\pi(t, \alpha_P(t))$ contains some additional b-mode factor [89]. Furthermoore, the factor $(1 + e^{-i\pi\alpha_P(t)})\,\Gamma(-\alpha_P(t))$ in eq. (II.82) is replaced by $\Gamma(1 - \alpha_P(t)/2)/\Gamma(\alpha_P(t)/2)$ so that the NS- or NSR-models do not contain a Pomeron tachyon pole at $\alpha_P(t) = 0$.

III.3.2. The singularities of the nonplanar orientable two-loop graph

Let us show how to rederive the multi-Pomeron asymptotics (III.82) in the case $M = 2$ starting from the integral representation (III.56) with (III.64). Now we have to handle the relevant two-loop automorphic functions without taking any reference to iterative one-loop structures. As we see from eq. (III.64) the moving Regge singularities in the t-channel arise from the singularities of the Neumann function $V_t = -N(z_1 z_4; z_5, z_8)$, either in the KN-variables z_i or in the $(3M-3)$ surface parameters (invariant points and multipliers) implicitely involved in N. Again, as $s \to i\infty$ a power behaviour arises from those critical points where $V_s = -N(z_8, z_1; z_4, z_5) = 0$. Because of eq. (III.57) this means $z_1 = z_8$ and/or $z_4 = z_5$. At fixed surface parameters this produces ordinary Regge behaviour s^α. For example, if $z_8 \to z_1$, then

$$\left.\begin{aligned} \frac{1}{2}\, sN(z_8, z_1; z_4, z_5) &\sim \frac{1}{2}\, s \,\mathrm{Re}\,[(z_8 - z_1)\, \Omega'_{z_4 z_5}(z_1)] \\ \frac{1}{2}\, tN(z_1, z_4; z_5, z_8) &\sim \frac{-t}{2} \log|z_1 - z_8| + \text{regular terms}. \end{aligned}\right\} \qquad \text{(III.89)}$$

But $z_1 = z_8$ or $z_4 = z_5$ can also be forced by a pinch of z_1, z_8 or z_4, z_5 of the isometric circles J_{T_i}, $J_{T_i}^{-1}$ of the projective transformations T_7 or T_3 (comp. eq. (III.63) and fig. 24) become tangent. Using the multiplier K_μ and the invariant points $\xi_\mu \eta_\mu$ of the projective transformation $T_\mu(z)$ (see eq. (III.43) this situation occurs in the so-called parabolic limit defined by $K_\mu \to 1$, $\xi_\mu \to \eta_\mu$ with $(\xi_\mu - \eta_\mu)/(1 - K_\mu) \to C_\mu$ [82]. In this limit the arcs (AB) or (CD) in fig. 24 as well as the isometric circles of the Jacobi transformed generators P_7, P_3 in the y-plane shrink to a point. If we denote by ω_7, ω_3 their respective multipliers we have also (due to eq. (III.44)) $\omega_2, \omega_7 = 0$. Thus the 'horizontal' cycles on the sphere with handles are shrinking (comp. fig. 21), i.e. there are handles vanishing and therefore the topology of the underlying Riemann surface changes. As is well-known, this behaviour leads to the appearance of the Pomeron singularity [76—78]. We only mention here that there is yet another possibility of changing the topology of the Riemann surface, namely $K_\mu \to 0$. Die to eq. (III.44) the radius of the associated iso-

metric circle then shrinks to zero so that the corresponding handle of fig. 21 vanishes by becoming infinitely thin. These singularities lead to a Regge cut behaviour [*77, 81*].
Let us now discuss further the iterated Pomeron singularity arising from the parabolic limit $\omega_3 \to 0$, $\omega_7 \to 0$. Using the identity

$$y_i - P_\alpha(y_j) = \frac{(y_i - \varrho_\alpha) - \omega_\alpha \dfrac{y_j - \varrho_\alpha}{y_j - \sigma_\alpha}(y_i - \sigma_\alpha)}{1 - \omega_\alpha \dfrac{y_i - \varrho_\alpha}{y_j - \sigma_\alpha}} \tag{III.90}$$

where $P_\alpha(y)$ is an arbitrary element of the automorphic group generated by P_3 and P_7 and expressing ω_α, ϱ_α, σ_α in terms of ω_i, ϱ_i, σ_i ($i = 3, 7$), one may now expand all the relevant functions appearing in the two-loop integrand (comp. eq. III.53–55) around the point $\omega_3 = \omega_7 = 0$. Choosing the Jacobi transformation in such a way that $\varrho_3 = \bar{\sigma}_3$, $\varrho_7 = \bar{\sigma}_7$ and using $\varepsilon_\mu \sim \sqrt{\omega_\mu}\,|\sigma_\mu - \varrho_\mu|$, ε_μ = radius of the isometric circle of P_μ)

$$\begin{aligned} y_\mu - \sigma_7 &= \varepsilon_7 e^{i\varphi_\mu} + O(\varepsilon_7^2) \qquad (\mu = 1, 8) \\ y_\mu - \sigma_3 &= \varepsilon_2 e^{i\varphi_\mu} + O(\varepsilon_3^2) \qquad (\mu = 4, 5) \end{aligned} \tag{III.91}$$

we obtain, for example, the following estimate of the third Abelian integral (III.54) [*80*]

$$\begin{aligned} \operatorname{Re} \omega(y_8, y_1; y_4, y_5) &= -8\varepsilon_3\varepsilon_7 \sin\frac{1}{2}(\varphi_8 - \varphi_1)\sin\frac{1}{2}(\varphi_4 - \varphi_5) \\ \times \operatorname{Re} &\left[\frac{\exp\left(\frac{i}{2}(\varphi_8 + \varphi_1 + \varphi_4 + \varphi_5)\right)}{(\sigma_7 - \sigma_3)^2} - \frac{\exp\left(\frac{i}{2}(\varphi_1 + \varphi_8 - \varphi_4 - \varphi_5)\right)}{(\sigma_7 - \bar{\sigma}_3)^2}\right] + o(\varepsilon_3 \cdot \varepsilon_7). \end{aligned} \tag{III.92}$$

Similar estimates can also be performed for the first Abelian integrals and the period matrix. Finally, the factors appearing in the integration measure given in terms of the original KN-variables z_i were Jacobi transformed by evaluating eq. (III.66) in the vicinity of the parabolic points. Integrating, we obtain a result as in (III.82) ($M = 2$) where the integral representation of the transition part π looks slightly different due to the other variables used. It is worth mentioning that π is closely related to the off-shell propagator of Cremmer and Scherk [*90*] introduced in their approach to dual current amplitudes.

III.4. Discussion of renormalization

As we have seen in the preceding sections, some expressions as the planar one-loop amplitude (III.3), the planar trajectory correction (III.72), the planar two-loop correction (III.75) and the Pomeron correction (III.84) have not well-defined integral representations due to divergencies of the one- or two-loop partition functions (and of some angular integrations). To give a physically sensible meaning to the unitarization program the corresponding integrals have to be regularized by subtracting out the infinite pieces in a manner which is compatible with duality, factorization and Regge behaviour. Now there arises naturally the question whether these infinite pieces lead in analogy to conventional field theory to a renormalization of the bare parameters a, α' and g of the dual theory. As the ω-integrand of the one-loop amplitude (III.3) is divergent as $f(\omega)^{E-D/2} \underset{\substack{\omega\to 1\\(q\to 0)}}{\sim} q^{-D-E/12}$ (comp. eqs. (III.8, 9), the number of counterterms depends crucially on the space time-dimension D. For example, in the case $D = 4$ one needs only one counterterm to make the q integral finite [*91*]. Depending on whether we have then an intercept value $a \neq 1$ or $a = 1$, respectively, the arbitrariness of this counterterm

turns out to be quite different. Thus, for $a \neq 1$ the counterterm can only be defined up to an arbitrary function $G(q)$ in the q-integral so that an arbitrary function of the external momenta (satisfying duality and Regge behaviour, etc.) is left free.
In the Virasoro case $a = 1$, however, the q-integration can be factorized out in the counterterm leaving us with one multiplicative constant only. As in this case gauges are working there arises no renormalization of the intercept of the bare trajectory $\alpha(t) = 1 + \alpha' t$ in close analogy to massless Yang-Mills field theory or quantum electrodynamics [*91*]. Instead, one obtains a renormalization of the bare coupling constant.
Let us now discuss in some more detail the interesting case of critical space-time dimension where there are no unitarity violating cuts of the dual Pomeron graph. As the integrand behaves now as $q^{-\alpha_P(0)-1} = q^{-3}$, the subtraction of one counterterm is not enough to make the q-integral finite. To investigate this case $D = D_{\text{crit}} = 26$ further, we shall use the important fact recognized first by GODDARD [*92*] that the planar loop may be viewed as a nonplanar one where all the particles on the internal boundary (comp. fig. 20) have zero four-momenta. This means that one can re-express the planar loop in terms of the emission of a Pomeron with four-momentum $p = 0$ into the vacuum (see fig. 27).

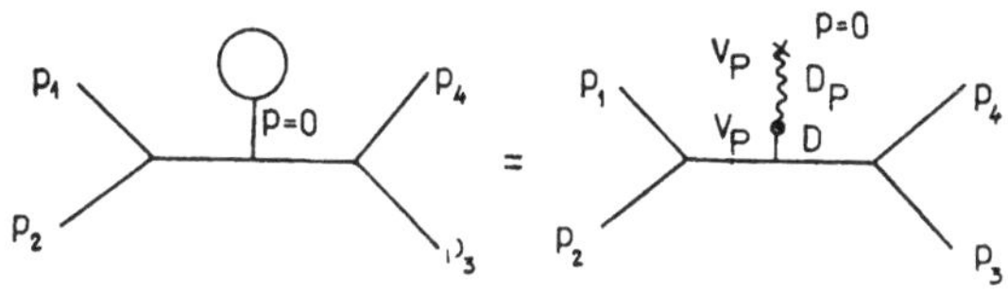

Fig. 27. Factorization of the planar one-loop graph making explicit the connection to a Pomeron transition into the vacuum

Thus, we have [*69*]

$$T_1^{\text{pl}}(s,t) = \lim_{p=\sum_i p_i \to 0} \langle 0 | \Sigma_T(p) | t; p_1, p_2, p_3, p_4 \rangle = \lim_{p=\sum_i p_i \to 0} \langle O | V_P{}^+ D_P V_P D | t; p_1, p_2, p_2, p_4 \rangle \tag{III.93}$$

where $|t; p_1, p_2, p_3, p_4\rangle$ is a tree state as defined in eq. (II.48) with $\sum_{i=1}^{4} p_i = p$ and V_P, D_P are the Pomeron vertex and propagator introduced in sect. III.2.3. Note that the above prescription of analytical continuation makes the $q^{-3-p^2/4}$-term finite and avoids also spurious singularities in the angular integrations of the loop as arising from the ordinary propagator $D(0)$. The only divergences appearing in eq. (III.93) due to the propagator $D_P(p)$ are then connected with the Pomeron tachyon ground state $|0\rangle_{c\bar{c}}$ and the so-called dilaton state $c_1{}^{\mu+} c_{1\mu}^{+} |0\rangle_{c\bar{c}}$ which have to be subtracted out of eq. (III.93). This yields

$$\begin{aligned} T_1^{\text{pl}}(s,t) = \lim_{p\to 0} g^4 \int dq\, q^{-\alpha_P(p^2)-1} \Big\{ f^{-24}(q^2) \Big(\prod_{i=2}^{4} \int d\Theta_i \Big) \\ \times \prod_{i<j} [\bar{\psi}(\Theta_j - \Theta_i)]^{-p_i p_j} - \Big(\prod_{i=2}^{4} \int d\Theta_i \Big) \prod_{i<j} [\sin(\Theta_j - \Theta_i)]^{-p_i p_j} \\ \times \Big[1 + q^2 (24 - \sum_{i,j} 2 p_i p_j \sin^2(\Theta_j - \Theta_i) \Big] \Big\} \\ + \lim_{p\to 0} \Big[\frac{g^4}{-\alpha_p(p^2)} \mathfrak{F}_4(p; p_1, p_2, p_3, p_4) + \frac{24 g^4}{2 - \alpha_p(p^2)} \mathfrak{F}_4(p; p_1, \dots, p_4) \\ - \frac{2g^4}{2 - \alpha_p(p^2)} \int \prod_{i=2}^{4} d\Theta_i \prod_{i<j} \sin(\Theta_j - \Theta_i)^{-p_i p_j} \sum_{i,j} p_i p_j \sin^2(\Theta_j - \Theta_i) \Big] \end{aligned} \tag{III.94}$$

where the function $\mathfrak{F}_4$ is defined in eq. (III.95). If we further use [69]

$$\lim_{p\to 0} \mathfrak{F}_4(p, p_1, \dots p_4) = \lim_{p=\sum_i p_i \to 0} \int \prod_{i=2}^{4} d\Theta_i \prod_{i<j} \sin (\Theta_j - \Theta_i)^{-p_i p_j} = 2\pi B_4(1, 2, 3, 4) \tag{III.95}$$

we see that the first two terms in the square bracket of eq. (III.94) contribute again to a coupling constant renormalization. Let us now take the asymptotic behaviour ($|s| \to \infty$, t fix) of the counterterm $C(s, t)$ defined by the last row of eq. (III.94). We obtain [93]

$$C(s, t) \sim g^2 \frac{\partial}{\partial l} \left[\left| \frac{s}{2} \right|^l \Gamma(-l)\, g^2 \tilde{\Sigma}(l) \right]_{l=\alpha(t)} \tag{III.96}$$

where

$$\tilde{\Sigma}\big(\alpha(t)\big) = \lim_{p\to 0} \frac{4t}{2 - \alpha_P(p^2)}\, B\left(\frac{1}{2}, \frac{1}{2}\right). \tag{III.97}$$

This result suggests the following renormalization prescription for the planar self-energy correction $\Sigma\big(t, \alpha(t)\big)$ defined in eq. (III.72)[20])

$$\Sigma\big(t, \alpha(t)\big) = \lim_{p\to 0} \int_0^1 dq q^{-\alpha_P(p^2)-1} \left[\left\{ f^{-24}(q^2) \int_0^\pi d\Theta [\overline{\psi}(\Theta)]^{2\alpha(t)-1-\alpha(p^2)} \left(-\frac{\partial^2}{\partial \Theta^2} \ln \overline{\psi}(\Theta) \right)^{\alpha(t)} \right\} - \{\dots\}|_{q=0} - q^2 \frac{\partial}{\partial(q^2)} \{\dots\}|_{q=0} \right] + \lim_{p\to 0} \frac{4t}{2 - \alpha_P(p^2)}\, B\left(\frac{1}{2}, \frac{1}{2}\right) \tag{III.98}$$

where we renounced to add once more the first counterterm since

$$\lim_{p\to 0} \int_0^\pi [\sin \Theta]^{-1-\alpha(p^2)} = \lim_{p\to 0} B\left(\frac{1}{2}, -\frac{1}{2}\alpha(p^2)\right) = 0. \tag{III.99}$$

We then find up to order g^4 the following trajectory correction

$$\alpha^{\text{new}}(t) = 1 + \alpha' \left(1 + \lim_{p\to 0} \frac{8\pi g^2 b}{2 - \alpha_P(p^2)}\right) t + g^2 \cdot b[\Sigma\big(t, \alpha(t)\big) - \tilde{\Sigma}\big(t, \alpha(t)\big)] + O(g^4). \tag{III.100}$$

In the case of the critical space time dimension one has thus besides a coupling constant renormalization an additional renormalization of the Regge slope α' in agreement with the results of ref. [94]. Analogous results have been obtained for the renormalized ϱ-trajectories of the NSR-model with $D = D_{\text{crit}} = 10$ [93], too.

One may now generate multi-loop amplitudes by iterating operatorial parts $\lim_{P\to 0} \langle 0 | \Sigma_T(p)$ and applying them to tree states. The main problem is here that one has not only to subtract tachyon and dilaton contributions of the single loops but also the contributions of higher multi-loop divergences (there are at least $2^M - 1$ singularities in a M-loop diagram

[20]) This expression for Σ differs by a factor $(-1)^{\alpha(t)}$ from eq. (III.72) because the Reggeons between the self-energy parts are nontwisted. In the following we include a relative weight factor $b \propto (2\alpha')^{3-D/2}$ in the expression III.98 to have the same mass dimension as the tree, $[m]^{4-D}$. Now g has the dimension$[m]^{(6-D)/2}$ and each propagator gets a factor α', $[\alpha'] = [m]^{-2}$.

due to divergences of the multiloop paritition function [*95*]). In fig. 28 we show the additional divergence of the planar and nonplanar two-loop trajectory correction due to a third Pomeron singularity emitted into the vacuum.
To subtract this new divergence one has to perform the factorizations shown in fig 28 defining a three-Pomeron vertex. This problem has not yet been solved. For the treatment of multi-loop divergencies (for $D \neq 26$) we refer the reader also to ref. [*96*].

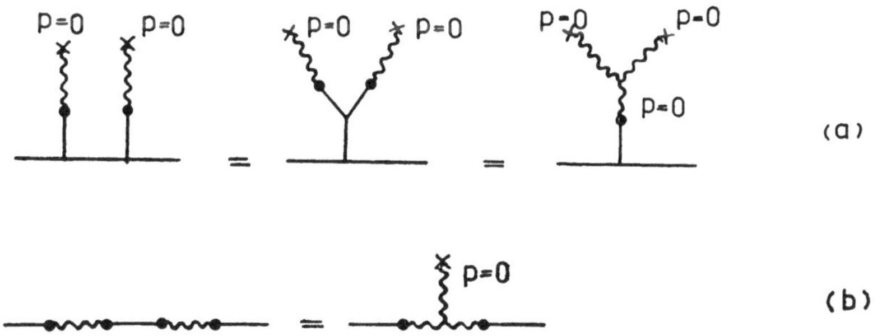

Fig. 28. Different factorized representations of the planar (a) and nonplanar (b) two-loop structures with a third-Pomeron

IV. Concluding Remarks

This review should give at least a rough inside into the high degree of selfconsistency and closure of the present-day dual theory, which has undoubtely enriched our understanding of hadron, may be even nonhadron dynamics considerably. Despite the huge amount of progress achieved during the last years there remain yet some important problems to be solved in future. It is widely acknowledged that the most severe one is related to the removal of the tachyon from the spectrum. There is some hope and even some prospecting work [*29*] that there could be used some spontaneous symmetry breaking for shifting the vacuum and thereby the mass spectrum. The second important defect consists in the unphysically large dimensionality of space-time as demanded primarily by the Pomeron to be a pole[21]). A possible way out of both the tachyon and dimensionality problem has quite recently been proposed in ref. [*97*]. Furthermore, it would be very important especially in the light of the recent developments in deep inelastic electron-hadron scattering to have dual current amplitudes. Though there have been many attempts to get dual currents [*90, 98*] none of these was completely satisfactory up to now. Finally, at the level of loop diagrams there remains the problem to renormalize the two-and higher loop diagrams in critical dimensions. The prerequisite for this probably are the removal of spurious states in multi-loops and the explicit factorization of the 'third' Pomeron singularity. It would then be very interesting to check the conjecture [*94*] that the dual renormalization programme generally leads to redefinitions of the coupling constant and the Regge slope parameter only.

One of us (H.-J.O.) thanks for the hospitality extended to him at the Theory Division of the Joint Institute für Nuclear Research, Dubna.

[21]) In the string formalism one sees a one-to-one correspondence between this 'critical' dimensionality (and intercept $a = 1$) to the demand that there exist only transverse modes of the string [*5*].

Appendix A

In this appendix we apply a method used in ref. [62] to calculate the asymptotic behaviour as $s \to i\infty$ of an integral of the following type

$$I(\alpha_1, \alpha_2, \alpha_3; s) = \int_0^1 dx \int_0^1 dy \int_0^1 dz x^{-\alpha_1-1} y^{-\alpha_2-1} z^{-\alpha_3-1} \Phi(x, y, z)\, e^{-sF(x,y,z)} \qquad (\alpha_1, \alpha_2, \alpha_3 > 0) \tag{A.1}$$

where near the point $x = y = z = 0$

$$F(x, y, z) = \{+xyzF_{xyz}(0, 0, 0) + (\text{higher orders in } x, y, z)\}. \tag{A.2}$$

Finally, at the end of the calculation we are interested in the case $\alpha_1 = \alpha_2 = \alpha_3 = \alpha$. First we have to determine the critical points of the integrand of eq. (A.1). On the boundary $x = 0$ we have $\partial^2 F(0, y, z)/\partial y\, \partial z = 0$ for all y, z, i.e. the whole boundary $x = 0$ is made up from critical points. The same holds also for the boundaries $y = 0$ and $z = 0$. The asymptotic behaviour of the integral (A.1) is then obtained by adding the contributions of the 'critical boundaries' $x = 0$ all y, z, $y = 0$ all x, z and $z = 0$ all x, y. Thus we may write

$$\begin{aligned}
\lim_{s\to i\infty} I(\alpha_1, \alpha_2, \alpha_3; s) &= \iint_0^1 dx\, dy x^{-\alpha_1-1} y^{-\alpha_2-1} \lim_{s\to i\infty} \int_0^1 dz z^{-\alpha_3-1} \Phi(x, y, z)\, e^{-sF(x,y,z)} \\
&+ \iint_0^1 dx\, dz x^{-\alpha_1-1} z^{-\alpha_3-1} \lim_{s\to i\infty} \int_0^1 dy y^{-\alpha_2-1} \Phi(x, y, z)\, e^{-sF(x,y,z)} \\
&+ \iint_0^1 dy\, dz y^{-\alpha_2-1} z^{-\alpha_3-1} \lim_{s\to i\infty} \int_0^1 dx x^{-\alpha_1-1} \Phi(x, y, z)\, e^{-sF(x,y,z)} \\
&\equiv I^{(1)}(\alpha_1, \alpha_2, \alpha_3; s) + I^{(2)}(\alpha_1, \alpha_2, \alpha_3; s) + I^{(3)}(\alpha_1, \alpha_2, \alpha_3; s).
\end{aligned} \tag{A.3}$$

Let us first perform the z-integration in the term $I^{(1)}(\alpha_1, \alpha_2, \alpha_3; s)$. If we add and subtract suitable terms to make the remaining x,y-integrations for $\alpha_1 = \alpha_2 = \alpha_3 = \alpha$ finite we obtain

$$\begin{aligned}
I^{(1)}(\alpha_1, \alpha_2, \alpha_3; s) \underset{s\to i\infty}{\sim} & s^{\alpha_3}\Gamma(-\alpha_3) \iint_0^1 dx\, dy x^{-\alpha_1-1} y^{-\alpha_2-1} (1 - O_x)(1 - O_y) F_z(x, y, 0)^{\alpha_3} \Phi(x, y, 0) \\
&+ \frac{s^{\alpha_3}\Gamma(-\alpha_3)}{\alpha_3 - \alpha_1} \int_0^1 dy y^{-\alpha_2-1} (1 - O_y) F_{xz}(0, y, 0)^{\alpha_3} \Phi(0, y, 0) \\
&+ \frac{s^{\alpha_3}\Gamma(-\alpha_3)}{\alpha_3 - \alpha_2} \int_0^1 dx x^{-\alpha_1-1} (1 - O_x) F_{yz}(x, 0, 0)^{\alpha_3} \Phi(x, 0, 0) \\
&+ \frac{s^{\alpha_3}\Gamma(-\alpha_3)}{(\alpha_3 - \alpha_2)(\alpha_3 - \alpha_1)} F_{xyz}(0, 0, 0)^{\alpha_3} \Phi(0, 0, 0)
\end{aligned}$$

$$\left(F_x = \frac{\partial F}{\partial x} \text{ etc.}\right). \tag{A.4}$$

Here we have introduced, for convenience, an operator $O_{x,y,z}$ acting in the following way

$$O_x F(x, y, z)^\alpha \Phi(x, y, z) = F_x(0, y, z)^\alpha \Phi(0, y, z) x^\alpha . \tag{A.5}$$

Similar expressions may also be derived for $I^{(2)}$ and $I^{(3)}$. If we now consider the limit $\alpha_1, \alpha_2, \alpha_3 \to \alpha$ the single and double poles of eq. (A.4) are cancelled by analogous terms of $I^{(2)}$ and $I^{(3)}$. We, finally, arrive at

$$I(\alpha, \alpha, \alpha; s) \underset{s \to i\infty}{\sim} s^\alpha \Gamma(-\alpha) \left\{ \int_0^1 \int dx\, dy (xy)^{-\alpha-1} (1 - O_x)(1 - O_y) F_z(x, y, 0)^\alpha \Phi(x, y, 0) \right.$$
$$\left. + \text{ analogous } (x,z)\text{- and } (y,z)\text{-integrals} \right\}$$
$$+ \frac{\partial}{\partial \alpha} \left[s^\alpha \Gamma(-\alpha) \left\{ \int_0^1 dx x^{-\nu-1} (1 - O_x) F_{yz}(x, 0, 0)^\alpha \Phi(x, 0, 0) \right.\right.$$
$$\left.\left.\left. + \text{ analogous } y\text{- and } z\text{-integrals} \right\} \right] \right|_{\nu=\alpha}$$
$$+ \frac{1}{2!} \frac{\partial^2}{\partial \alpha^2} [s^\alpha \Gamma(-\alpha) F_{xyz}(0, 0, 0)^\alpha \Phi(0, 0, 0)] \tag{A.6}$$

Note that the last term in eq. (A.6) may be thought of as being obtained by expanding the integrand around the corner $x = y = z = 0$.

Appendix B

In this appendix we collect the iteration equations for the E_M, X_M and F_M matrices of the M-loop self-energy operator $\sum_M{}^{(T)}$ of eq. (III.39). We have (for notations comp. eq. (III.12))

$$E_M{}^{(l)} = E_{M-1}^{(l)} + X_{M-1}\{z_{M-1}\} E(M) \Phi_M\{z_{M-1}\} \tilde{X}_{M-1}$$

$$E_M{}^{(r)} = E(M) + \tilde{X}(M) \{z_{M-1}\} E_{M-1}^{(r)}\{z_{M-1}\} \tilde{\Phi}_M X(M)$$

$$X_M = X_{M-1}\{z_{M-1}\} \tilde{\Phi}_M X(M)$$

$$F_M{}^{(l)} = F_{M-1}^{(l)} + [F^{(l)}(M) - F_{M-1}^{(r)}\{z_{M-1}\} E(M)] \Phi_M\{z_{M-1}\} X_{M-1}$$

$$F_M{}^{(r)} = F(M) + [F_{M-1}^{(r)} - F(M) \{z_{M-1}\} E_{M-1}^{(r)}] \{z_{M-1}\} \tilde{\Phi}_M X(M)$$

$$\Phi_M = [1 - \{z_{M-1}\} E_{M-1}^{(r)}\{z_{M-1}\} E(M)]^{-1}$$

$$\varphi_M = \varphi_{M-1} + \varphi(M) + \frac{1}{2} (1| F_{M-1}^{(r)}\{z_{M-1}\} E(M) \Phi_M\{z_{M-1}\} \tilde{F}_{M-1}^{(r)} |1)$$
$$+ \frac{1}{2} (1| F(M) \Phi_M\{z_{M-1}\} E_{M-1}^{(r)}\{z_{M-1}\} \tilde{F}(M) |1) - (1| F(M) \Phi_M\{z_{M-1}\} \tilde{F}_{M-1}^{(r)} |1). \tag{B.1}$$

Here $\tilde{X}$ means' transpose of X' etc., $E(M)$ and $F(M)$ are the one-loop matrices depending on the variables of the M'th $\sum^{(T)}$ operator. In the case of $\sum^{T}$ we have, for example,

$$E_{mn}^{(l,r)} = \frac{1}{2}\frac{\sqrt{mn}}{m!\,n!}\left(\frac{\partial}{\partial x}\right)^m\left(\frac{\partial}{\partial y}\right)^n \ln\left[\frac{\overline{\psi}(\pi(x-y))}{\pi(x-y)}\right]_{x=y=0}$$

$$\underset{q\to 0}{\sim} -\frac{1}{2}\frac{\sqrt{mn}}{m!\,n!}\left(\frac{\partial}{\partial x}\right)^m\left(\frac{\partial}{\partial y}\right)^n \ln\left[\frac{\sin\pi(x-y)}{\pi(x-y)}\right]\Bigg|_{x=y=0} \equiv F_{mn}^0 \qquad \text{(B.2)}$$

$$X_{mn} = -\frac{\sqrt{mn}}{m!\,n!}\left(\frac{\partial}{\partial x}\right)^m\left(\frac{\partial}{\partial y}\right)^n \ln\left[\overline{\psi}_T(\Theta + \pi(x+y))\right]_{x=y=0}$$

$$\underset{q\to 0}{\sim} q\,\frac{\sqrt{mn}}{m!\,m!}\left[e^{2i\Theta}(2\pi i)^{m+n} + e^{2i\Theta}(-2\pi i)^{m+n}\right] \equiv qX_{mn}^0(\Theta) \qquad \text{(B.3)}$$

$$\{(1|\,F^{(l,r)}\}_m = -\frac{\sqrt{m}}{m!}\left(\frac{\partial}{\partial x}\right)^m \ln\left[\frac{\overline{\psi}_T(\Theta+\pi x)\,\pi x}{\overline{\psi}(\pi x)}\right]_{x=0}$$

$$\underset{q\to 0}{\sim} \frac{\sqrt{m}}{m!}\left(\frac{\partial}{\partial x}\right)^m \ln\left[\frac{\sin\pi x}{\pi x}\right]_{x=0} \equiv \{(1|\,F^0\}_m. \qquad \text{(B.4)}$$

Let us expand the nonplanar M-loop functions around the corner $q_1 = q_2 = \cdots = q_M = 0$. Taking into account eqs. (B.2 – B.4) one obtains from eq. (B.1)

$$F_M^{(l)} \sim F_M^{(r)} = F^0 + O(q_i)$$

$$E_M^{(l)} \sim E_M^{(r)} = E^0 + O(q_i^2)$$

$$\Phi_M = [1 - \{z_{M-1}\}\,E^0\{z_{M-1}\}\,E^0]^{-1} + O(q_i) \equiv \Phi_M{}^0(z_{M-1}) + O(q_i)$$

$$\varphi_M = -\sum_{i=1}^{M-1}(1|\,F_0\,\frac{1}{1+\{z_i\}\,E_0}\,\{z_i\}\,F_0\,|1) + O(q_i) \equiv \sum_{i=1}^{M-1}F(z_i) + O(q_i). \qquad \text{(B.5)}$$

Finally, the integration measure $d\tau_M$ is defined by the iteration formula

$$d\tau_M = d\tau_{M-1}\,d\tau(M)\,dz_{M-1}(z_{M-1})^{-\alpha(t)-1}\,[\det\Phi_M{}^{-1}]^{-D/2} \qquad \text{(B.6)}$$

References

[1] G. VENEZIANO, Nuovo Cimento **57 A**, 190 (1968).

[2] J. SCHERK, Nuclear Phys. **B31**, 222 (1971); A. NEVEU, J. SCHERK, Nuclear Phys. **B36**, 155 (1972).

[3] A. NEVEU and J. H. SCHWARZ, Nuclear Phys. **B31**, 86 (1971); A. NEVEU, J. H. SCHWARZ and C. B. THORN, Phys. Letters **35B**, 529 (1971).

[4] P. RAMOND, Phys. Rev. **D3**, 2415 (1971).

[5] P. GODDARD, J. GOLDSTONE, C. REBBI and C. B. THORN, Nuclear Phys. **56B**, 109 (1973).

[6] S. MANDELSTAM, Nuclear Phys. **B64**, 205 (1973); E. CREMMER and J. L. GERVAIS, Nuclear Phys. **B76**, 209 (1974). M. KAKU and K. KIKKAWA, Phys. Rev. **D10**, 1110 (1974); C. REBBI, Phys. Reports **12C**, 1 (1974).

[7] Y. IWASAKI and K. KIKKAWA, Phys. Rev. **D8**, 440 (1973); S. MANDELSTAM, Nuclear Phys. **B69**, 77 (1974).

[8] V. ALESSANDRINI, D. AMATI, M. LE BELLAC and D. OLIVE, Phys. Reports **1C**, 269 (1971).

[9] J. H. Schwarz, Phys. Reports **8C**, 4 (1973).
[10] S. Mandelstam, Phys. Reports **13**, 259 (1974).
[11] J. Scherk, "An Introduction to the Theory of Dual Models and Strings", prepr. (1974).
[12] P. Frampton, Dual Models (Wiley-Interscience Publication) M. Jacob (ed.), "Dual Theory", (North Holland, 1974).
[13] G. F. Chew, The Analytic S-matrix, W. A. Benjamin (1966); N. N. Bogoljubov and D. V. Shirkov, Introduction into the Theory of Quantum Fields, Moscow (1973); P. D. B. Collins and E. J. Squires, Regge Poles in Particle Physics, Berlin/Heidelberg/New York (1968).
[14] M. A. Virasoro, Phys. Rev. **177**, 2309 (1969); J. A. Shapiro, Phys. Letters **33B**, 361 (1970); M. Yoshimura, Phys. Letters **34B**, 79 (1971).
[15] H. Harari, Phys. Rev. Letters **22**, 562 (1969); J. L. Rosner, Phys. Rev. Letters **22**, 689 (1969).
[16] H. B. Nielsen, NORDITA prepr. (1971); D. Fairlie and H. B. Nielsen, Nuclear Phys. **20B**, 637 (1970).
[17] K. Bardakci and H. Ruegg, Phys. Letters **28B**, 342 (1968); C. G. Goebel and B. Sakita, Phys. Rev. Letters **22**, 257 (1969); Chan Hong-Mo and T. S. Tsun, Phys. Letters **28B**, 425 (1969).
[18] Z. Koba and H. B. Nielsen, Nuclear Phys. **B10**, 633 (1969).
[19] C. Lovelace, Phys. Letters **32B**, 490 (1970).
[20] S. Fubini, D. Gordon and G. Veneziano, Phys. Letters **29B**, 679 (1969); Y. Nambu, Proc. Intern. Conf. on symmetries and quark models Wayne State Univ., 1969.
[21] S. Fubini and G. Veneziano, Nuovo Cimento **67A**, 29 (1970).
[22] F. Gliozzi, Nuov. Cimento Letters **2**, 846 (1969).
[23] M. A. Virasoro, Phys. Rev. **D1**, 2933 (1970).
[24] S. Fubini and G. Veneziano, Nuovo Cimento **64A**, 811 (1969); K. Bardakci and S. Mandelstam, Phys. Rev. **184**, 1640 (1969).
[25] R. Hagedorn, Nuovo Cimento **56A**, 1027 (1968); S. Frautschi, Phys. Rev. **D3**, 2821 (1971).
[26] M. J. Gorenstein, V. A. Miransky, V. P. Shelest, G. M. Zinovjev and H. Satz, Kiev-Bielefeld preprint 73/12 (1973).
[27] Silvan, S. Schweber, An Introduction to Relativistic Quantum Field Theory, New York, Tokyo (1964).
[28] E. del Giudice, P. Di Vecchia and S. Fubini, Ann. Phys. (N.Y.) **70**, 378 (1972); P. Goddard and C. B. Thorn, Phys. Letters **40B**, 235 (1972); R. C. Brower, Phys. Rev. **D6**, 1655 (1972).
[29] K. Bardakci, Nuclear Phys. **B68**, 331 (1974); **B70**, 397 (1974); D. V. Volkov, A. A. Sheltuchin, A. J. Pashnev, Jadern. Fiz. **21**, 611 (1975); **21**, 1104 (1975).
[30] L. Caneschi, A. Schwimmer and G. Veneziano, Phys. Letters **30B**, 351 (1969).
[31] D. Olive, Nuovo Cimento **3A**, 399 (1971).
[32] Chan Hong-Mo and J. Paton, Nuclear Phys. **B10**, 519 (1969).
[33] J. Baacke, Nuovo Cimento **69A**, 297 (1970).
[34] C. Lovelace, Phys. Letters **28B**, 265 (1968). J. A. Shapiro, Phys. Rev. **179**, 1345 (1969).
[35] A. Neveu and J.-H. Schwarz, Phys. Rev. **D4**, 1109 (1971). C. B. Thorn, Phys. Rev. **D4**, 1112 (1971).
[36] E. F. Corrigan, P. Goddard, Nuclear Phys. **B68**, 189 (1974).
[37] E. Corrigan and D. Olive, Nuovo Cimento **11A**, 749 (1972).
[38] L. Brink, D. Olive, C. Rebbi and J. Scherk, Phys. Letters **45B**, 379 (1973).
[39] E. Corrigan, P. Goddard, D. Olive and R. A. Smith, Nuclear Phys. **B67**, 477 (1973).
[40] D. Olive and J. Scherk, Nuclear Phys. **B64**, 334 (1973); D. Bruce, E. Corrigan, D. Olive, Nuclear Phys. **B95**, 427 (1975).
[41] B. Petersson and N. A. Törnqvist, Nuclear Phys. **B13**, 629 (1969).
[42] Chan Hong-Mo, R. O. Raitio, G. H. Thomas and N. A. Törnqvist, Nuclear Phys. **B19**, 173 (1970).
[43] J. Bender, H. G. Dosch, V. F. Müller and H. J. Rothe Z. Phys. **237**, 107 (1970); A. C. Hirschfeld, J. G. Körner and M. G. Schmidt, Z. Phys. **252**, 255 (1972); P. Aurenche, Phys. Rev. **D9**, 1514 (1974) and **D9**, 1531 (1974).
[44] L. Becker, D. Ebert, H. J. Schreiber and G. Weigt, Berlin-Zeuthen prepr. PHE 6 (1975).
[45] L. Becker, E. Matthäus, H. J. Schreiber and H. Weigt, Berlin-Zeuthen prepr. PHE 5 (1975).
[46] H. A. Mueller, Phys. Rev. **D2**, 2963 (1970).

[47] M. A. Virasoro, Phys. Rev. D3, 2834 (1971).
[48] D. Gordon and G. Veneziano, Phys. Rev. D3, 2116 (1971).
[49] C. E. de Tar, K. Kang, C. I. Tan and J. H. Weis, Phys. Rev. D4, 425 (1971).
[50] K. J. Biebl, D. Bebel and D. Ebert, Berlin-Zeuthen, Prepr. 71-9 (1971), Fortschr. Phys. 20, 541 (1972).
[51] K. Huang and G. Segre, Phys. Rev. Letters 27, 1095 (1971).
[52] D. Ebert, Ann. Phys. (Leipzig) 29, 137 (1973).
[53] D. Bebel, K. J. Biebl, D. Ebert and H. J. Otto, Fortschr. Phys. 20, 555 (1972); D. Bebel, K. J. Biebl and H. Paul, Nuovo Cimento Letters 8, 85 (1973); D. Bebel and U. M. Wambach, Nuovo Cimento 20A, 307 (1974); G. H. Thomas, Phys. Rev. D5, 2212 (1971).
[54] C. L. Jen, K. Kang, S. Shen and C. I. Tan, Phys. Rev. Letters 27, 458 (1971) and Ann. Phys. 72, 458 (1972); S. Sakai, Progr. Theor. Phys. 47, 260 (1972); B. Hasslacher, C. S. Hsue and D. K. Sinclair, Phys. Rev. D4, 3089 (1971), see also prepr. of ref. [50].
[55] K. Kang and P. Shen, Phys. Rev. D7, 164 (1973); S. Fenster and J. L. Uretsky, Phys. Rev. D7, 2143 (1973).
[56] V. Alessandrini and D. Amati, Nuovo Cimento 13A, 663 (1973); D. Ebert and H. J. Otto, Nuovo Cimento 22A, 25 (1974); L. Masperi and C. Rebbi, Nuovo Cimento 13A, 689 (1973).
[57] D. Amati, C. Bouchiat, J. L. Gervais, Letters Nuovo Cimento 2, 399, (1969), K. Bardakci, M. B. Halpern, J. A. Shapiro, Phys. Rev. D185, 1910 (1969).
[58] D. J. Gross, A. Neveu, J. Scherk and J. H. Schwarz, Phys. Rev. D2, 697 (1970).
[59] R. C. Brower and C. B. Thorn, Nuclear Phys. B31, 163 (1971).
[60] L. Brink and D. Olive, CERN Prepr. TH 1773 (1973); L. Brink and D. Olive, Nuclear Phys. B58, 237 (1973); P. V. Collins and K. A. Friedman, Nuovo Cimento 28A, 475 (1975).
[61] L. Brink and D. Olive, Nuclear Phys. B56, 253 (1973).
[62] V. Alessandrini, D. Amati and B. Morel, Nuovo Cimento 7A, 797 (1972).
[63] H. Harari, Phys. Rev. Letters 20, 1395 (1968); P. G. O. Freund, Phys. Rev. Letters 20, 235 (1968).
[64] C. Lovelace, Phys. Letters 34B, 500 (1971).
[65] M. Froissart, Phys. Rev. 123, 1053 (1961).
[66] T. H. Burnett et al., Phys. Letters 32B, 115 (1970).
[67] D. J. Gross and J. H. Schwarz, Nuclear Phys. B23, 333 (1970).
[68] S. Sakai, Progr. Theor. Phys. 46, 590 (1971).
[69] E. Cremmer and J. Scherk, Nuclear Phys. B50, 222 (1972).
[70] P. Goddard and R. E. Waltz, Nuclear Phys. B34, 99 (1971); J. O. Winnberg, Nuclear Phys. B94, 205 (1975).
[71] L. Clavelli and J. A. Shapiro, Nuclear Phys. B57, 490 (1973).
[72] L. Brink and D. B. Fairlie, Nuclear Phys. B74, 321 (1974).
[73] D. Ebert and H. J. Otto, Ann. Phys. 32, 47 (1975).
[74] M. Ademollo, A. D'Adda, R. D'Auria, E. Napolitano, P. Di Vecchia, F. Gliozzi and S. Sciuto, Nuclear Phys. B77, 189 (1974).
[75] H. Dorn, E. Ebert and H. J. Otto, Acta Phys. Polonica B6, 599 (1975).
[76] C. Lovelace, Phys. Letters 32B, 703 (1970).
[77] V. Alessandrini, Nuovo Cimento 2A, 321 (1971).
[78] V. Alessandrini and D. Amati, Nuovo Cimento 4A, 793 (1971).
[79] W. Burnside, Proc. London Math. Soc. 23, 49 (1891).
[80] M. Schiffer and D. Spencer, Functionals on Finite Riemann Surface (Princeton, 1954).
[81] H. Dorn and D. Ebert, Nuclear Phys. B72, 228 (1974).
[82] L. Ford, Automorphic Functions (Chelsea, 1951).
[83] C. Montonen, Nuovo Cimento 19A, 69 (1974); D. B. Fairlie and D. Martin, Univ. of Durham prepr. 1973.
[84] D. S. Jones and M. Cline, J. Math. Phys. 37, 1 (1958).
[85] A. Neveu and J. Scherk, Phys. Rev. D1, 2355 (1970); H. J. Kaiser, F. Kaschluhn and E. Wieczorek, Berlin-Zeuthen, prepr. 71–3 (1971); H. Dorn and H. J. Kaiser, Acta Phys. Polonica B6, 17 (1975).
[86] A. D. Karpf and H. J. Liehl, Nuclear Phys. B74, 459 (1974); A. D. Karpf, Acta Phys. Austriaca, Suppl. XIII, 665 (1974).
[87] R. J. Eden, P. V. Landshoff, D. Olive and J. C. Polkinghorne: The Analytic S-matrix (Cambridge Univ. Press, 1966).

[88] C. Lovelace, Phys. Letters **40B**, 551 (1972); B. Morel and M. Quiros, Nuclear Phys. **B55**, 266 (1973).
[89] Hintze, W., Thesis 1975, Humboldt Univ. Berlin.
[90] E. Cremmer and J. Scherk, Nuclear Phys. **B48**, 29 (1972).
[91] A. Neveu and J. Scherk, Nuclear Phys. **B36**, 317 (1972).
[92] P. Goddard, Nuovo Cimento **4A**, 349 (1971).
[93] D. Ebert, H. J. Otto and V. Pervushin (to be published).
[94] I. A. Shapiro, Rutgers Univ. prepr. (1975); M. Ademollo, A. D'Adda, R. D'Autria, F. Gliozzi, E. Napolitano, S. Scinto, P. di Vecchia, Nuclear Phys. **B94**, 221 (1975).
[95] M. Kaku and J. Scherk, Phys. Rev. **D3**, 430 (1971).
[96] E. Cremmer and A. Neveu, Phys. Letters **33B**, 356 (1970).
[97] J. Scherk and J. H. Schwarz, Phys. Letters **57B**, (1975); E. Cremmer and J. Scherk, "Dual Models in Four Dimensions with Internal Symmetry, prepr. Oct. 1975.
[98] I. T. Drummond, Nuclear Phys. **B35**, 269 (1971); J. H. Schwarz and C. C. Wu, Nuclear Phys. **B 72**, 397 (1974); E. Corrigan and D. B. Fairlie, Nuclear Phys. **B91**, 527 (1975).

INHALT

Die „FORTSCHRITTE DER PHYSIK“ dienen der Publikation zusammenfassender Berichte über aktuelle Teilgebiete der Physik. Neben Originalberichten in deutscher oder englischer Sprache werden deutsche Übersetzungen wichtiger fremdsprachiger Berichte abgedruckt.

Aus dem Inhalt der folgenden Hefte:

H. MATSUMOTO, M. TACHIKI, H. UMEZAWA: The Boson Method for Anisotropic Superconductors

J. A. SWIECA: Solitons and Confinement

B. V. R. CHOWDARI, S. RADHAKRISHNA: Radiation Damage Products in Some Ionic Crystals

M. S. MARINOV, V. S. POPOV: Electron-Positron Pair Creation from Vacuum Induced by Variable Electric Field

E. W. MIELKE: Quantenfeldtheorie im de Sitter-Raum

H. HÜBEL: Kernstrukturaussagen magnetischer Momente von Hochspinzuständen

Manuskripte werden nach Vereinbarung mit dem Autor von der Schriftleitung entgegengenommen.

Bei Fragen zur Produktsicherheit wenden Sie sich bitte an:
If you have any questions regarding product safety,
please contact:

Walter de Gruyter GmbH
Genthiner Straße 13
10785 Berlin
productsafety@degruyterbrill.com